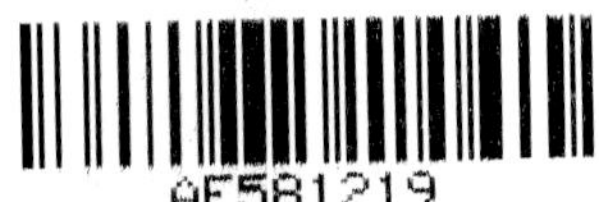

IMAGES
of America

WASHINGTON COUNTY'S ALUMINUM INDUSTRY

On the cover: The press room at the West Bend Aluminum Company is seen photographed around 1950. These huge machines are responsible for giving the basic shape to pots, pans, coffeemakers, and other kitchen utensils. (Courtesy of the Washington County Historical Society collections.)

IMAGES
of America

WASHINGTON COUNTY'S ALUMINUM INDUSTRY

Janean Mollet-Van Beckum and the
Washington County Historical Society

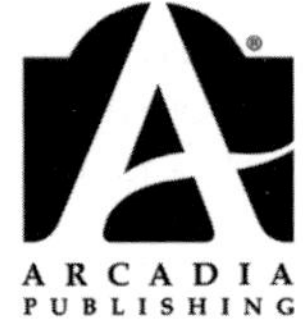

ISBN 978-0-7385-6044-1

Published by Arcadia Publishing
Charleston SC, Chicago IL, Portsmouth NH, San Francisco CA

Printed in the United States of America

Library of Congress Control Number: 2009925453

For all general information contact Arcadia Publishing at:
Telephone 843-853-2070
Fax 843-853-0044
E-mail sales@arcadiapublishing.com
For customer service and orders:
Toll-Free 1-888-313-2665

Visit us on the Internet at www.arcadiapublishing.com

Contents

Acknowledgments

This book would not have been possible without the dedication of the men and women of the aluminum and cookware industry in Washington County. The commitment and pride in the companies they worked for was key to their success. Thank you to the volunteers who have worked countless hours cataloging, filing, identifying, researching, and caring for this collection. We extend our sincere gratitude to Regal Ware Inc. and the Reigle family who have been invaluable with their support and information. We also appreciate Focus Products Group LLC for its support in this project. All images in this volume are from the collections of the Washington County Historical Society, Regal Ware Inc., and Focus Products Group LLC.

Introduction

Lightweight, economical, even heating, and easy to clean, aluminum was instrumental in the creation of the modern cookware industry. One of the most common elements on the planet, it is notoriously difficult to extract from surrounding materials. Aluminum was not used commercially until 1854, when the first extraction process was perfected. In 1850, aluminum cost $17 per pound. By the early 1900s, better extraction methods dropped the price to 34¢ per pound, making the metal available for mass production.

By 1920, Wisconsin became the center of the aluminum cookware industry, holding over 50 percent of the nation's production. Entrepreneurs established companies in areas where industry was needed to revive communities. Many of these, like West Bend and Kewaskum, offered less competition for skilled workers, capital, energy sources, and large markets nearby.

The aluminum industry in Washington County was a main employer in the 20th century. Manufacturing competitors, West Bend Aluminum Company and what began as the Kewaskum Aluminum Company, later to be Regal Ware Inc., both began production early in the century. Both have become well-known names worldwide and Regal Ware Inc. continues to be headquartered and manufacture goods in the county.

In 1911, a group of men headed by B. C. Ziegler organized the West Bend Aluminum Company. They had been looking for an industry to replace a leather goods company that was destroyed by fire. Ziegler had become interested in the industry after a trip to Two Rivers, where the successful Aluminum Goods Manufacturing Company was located. It was a stroke of good luck that the Aluminum Goods Company dismissed Carl and Robert Wentorf. The brothers were experienced tool and die makers and versed in the aluminum manufacturing process. Ziegler hired them, knowing they would be an asset to the young company. The first factory was housed in an old button factory and contained one draw press able to make about 15 different utensils. By 1914, the company expanded to a new facility.

During the Great Depression, West Bend Aluminum Company was able to keep many employees at work. Partly due to their efforts, the unemployment rate in the area was only 4 percent. Although small, the company did show a profit in 1932. It was also during the Depression that the company first manufactured copper products and in 1933 hired new workers.

Over the years, the West Bend Aluminum Company became a household name. During World War II, the company received six Navy E Awards. After World War II, the company purchased the former Kissel Motor Car Company plant in Hartford and began the outboard motor division. In 1957, West Bend Aluminum Company constructed plant facilities in Canada. By 1960, the company manufactured over 450 different products and employed over 2,000 people.

Over the years, the company continued to expand its holdings as well as its markets. In 1961, West Bend dropped aluminum from its name in accordance with the increased variety of products the company manufactured. Reaching into Europe, the Middle East, and Asia, the West Bend Company name became well-known internationally. In 1968, the company merged with Rexall Drug and Chemical Company, which later became Dart Industries. This began a long list of mergers and transfers for the company. In 1980, Dart Industries merged with Kraft Inc. and the West Bend Company became part of the newly formed Dart and Kraft. In 1996,

Dart and Kraft disbanded and the West Bend Company became part of Premark International, a corporation split from the former Dart and Kraft corporation. In 1999, Premark merged with Illinois Tool Works and the West Bend Company became a business unit of Illinois Tool Works. In 2002, Regal Ware Inc. acquired certain assets of the West Bend Company.

The company that became Regal Ware Inc. began as the Kewaskum Aluminum Company. In 1919, the Rosenhiemer Malt and Grain Company of Kewaskum had fallen on hard times due to Prohibition. Owner Adolph J. Rosenhiemer, seeing the great success of the neighboring West Bend Aluminum Company, decided to found his own aluminum-manufacturing corporation. After only 18 months in production, the company received the Good Housekeeping Seal.

After the death of Adolph Rosenhiemer, the company was sold to J. O. Reigle and associates in 1945. Knowledgeable in the aluminum production industry, Reigle renamed the company Kewaskum Utensil Company and shortly after renamed it Regal Ware Inc. Unlike the diversification of products from the West Bend Company, Regal Ware Inc. focused on the aluminum cookware industry and today still produces aluminum and stainless steel cookware as its main product. The company also expanded to become a worldwide name and in 2002 purchased West Bend Company.

In 2003, Focus Products Group LLC acquired the business and operations of the West Bend Company Retail Housewares Division from Regal Ware Inc. This acquisition included the small electrical appliance and timer businesses of the former West Bend Company. This enabled Regal Ware Inc. to focus on its core business of high-quality cookware manufacturing under both Regal Ware and West Bend brand names. In 2007, Focus Products Group acquired all the assets of the commercial business of Regal Ware Inc. This included the commercial coffee urns, commercial cookware, and other commercial product businesses.

Several factors made these two companies so successful. Foremost was the vision of the founders. In both cases, the founding fathers envisioned hard work while treating their employees with respect. They shared the companies' successes with their employees. An emphasis on quality and pride in workmanship were the staples that pushed these companies to the forefront of their industry. The ability to ship raw materials to the factory and finished products to the large markets of Milwaukee and Chicago via major rail routes resulted in fast turnaround, low prices, and happy customers.

Marketing genius and a bit of luck were also serendipitous to success. One example was the Waterless Cooker introduced by West Bend Aluminum Company in 1921. Because of the instruction needed to learn how to use it, the product did not sell well and was taken out of production in 1924. Other companies soon took up similar designs, and by the 1930s, the cooker was back in production and was sold by West Bend representatives though in-home demonstrations. It was products like the Waterless Cooker that made West Bend a household name. The other companies had made the investment of education, and West Bend reaped the benefits by outselling competitors.

The history of the companies would be remiss without a discussion of their broader influence. Both companies had a substantial affect on American consumerism. Innovations in cookware, such as Teflon-coated pans, were a housewife's dream on par with washing machines or vacuum cleaners. Advertising ease of use and fast preparations, the "ideal" housewife image of the 1950s was created. Product design both reflected and influenced the consumers' social values. For example, the 1950s popularity of dinner parties can be seen in a plethora of fashionable serving dishes and high capacity coffee urns. The marketing campaign told consumers that to entertain one needed to have nice-looking and high-performance products, thus influencing purchasing and idealization of a "good" housewife.

West Bend Company and Regal Ware Inc. are two of the best-known makers of cookware and small appliances around the world. Their ability to survive economic depression, war, and changing times are three reasons for their success and made them excellent examples of the American dream come to life. With deep roots in Washington County, they were able to spread wide nets capturing loyalty around the globe.

One

The Early Years 1911–1935

The earliest aluminum goods were novelty items such as combs, key fobs, and cigar cases. It was soon realized that aluminum served as an excellent material for cookware as well. These first utensils were practical. Design and appearance were strictly for functional use rather than decoration.

Beginning in the 1920s, market research allowed companies to design new technologies intended for the emancipated housewife and the "leisure living" of the era. As electric service was established in more locations, electric appliances were developed.

The Great Depression prompted aluminum companies to become more creative in their advertising and design as the competition for hard earned dollars increased. Specialty items and giftware sets helped drive companies like West Bend and Kewaskum Aluminum through the Depression. These hard times also saw a switch in the wants of buyers. Those who could afford luxury goods looked to food service products and personal status goods rather than food preparation items. Items such as cigar cases and serving trays are examples of unnecessary items the wealthy purchased to show their economic stability.

How Sheet Aluminum is Made

Aluminum is obtained from a mineral known as bauxite, one of the most abundant of all the metallic elements found in the earth's crust. It is mined in North and South America as well as in Europe and Asia. Bauxite occurs in a wide variety of colors and textures — it may be as hard as rock or as soft as clay — it may be pink, yellow, red or almost white, or a combination of these colors.

Mining Bauxite, the mineral from which aluminum is made.

In the first step, the aluminum oxide must be separated from impurities such as the oxides of iron and silicon which are also found in the bauxite. By a refining process, the aluminum oxide is then converted to a white powder called "alumina".

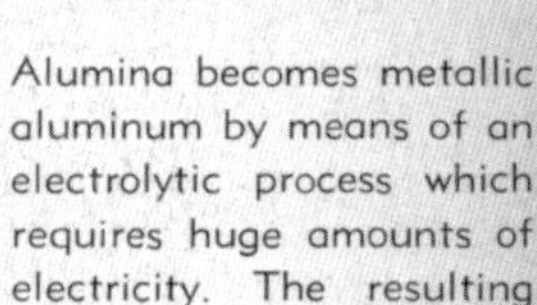

Alumina becomes metallic aluminum by means of an electrolytic process which requires huge amounts of electricity. The resulting molten metal is cast into pigs weighing approximately 50 pounds each.

onverting aluminum oxide to white powder called "alumina."

The aluminum pigs are remelted and cast into ingot which is rolled or formed into thick slabs. Each slab goes through a rolling mill where it is continuously rolled, annealed and rerolled, until it becomes a dense, hard sheet of aluminum. Cooking utensils and other products are then made from this metal which is free from pores and impurities.

Molten aluminum metal being cast into pigs.

As a recap of materials necessary to make one pound of aluminum:

4 pounds of Bauxite are required to make 2 lbs. alumina. 2 pounds of alumina required to make 1 lb. of aluminum.

Also 12 kilowatt hours of electricity and ¾ lb. of carbon electrode were consumed in making 1 lb. of aluminum.

If we take into account all materials from mining the Bauxite to the finished aluminum, we find it takes 9 pounds of raw materials to make 1 lb. of aluminum.

Aluminum is continuously rolled, annealed and rerolled into dense, hard sheet.

These pages from an early West Bend booklet explain the aluminum extraction process. Aluminum is obtained from a sedimentary rock called bauxite and is one of the most abundant metals in the earth's crust. The photograph in the upper left shows miners extracting bauxite from a mine around 1909. After the aluminum is separated from the other elements in the bauxite, it is refined into a white powder called alumina. The alumina is then converted to metallic aluminum by the Hall-Heroult process, which uses electricity to separate out pure aluminum. The aluminum is then melted and cast into pigs, from which it can be formed into many shapes. The photograph in the bottom right shows the finished aluminum being rolled into sheets. This form is the type most commonly used by the aluminum industry for use in drawn utensils.

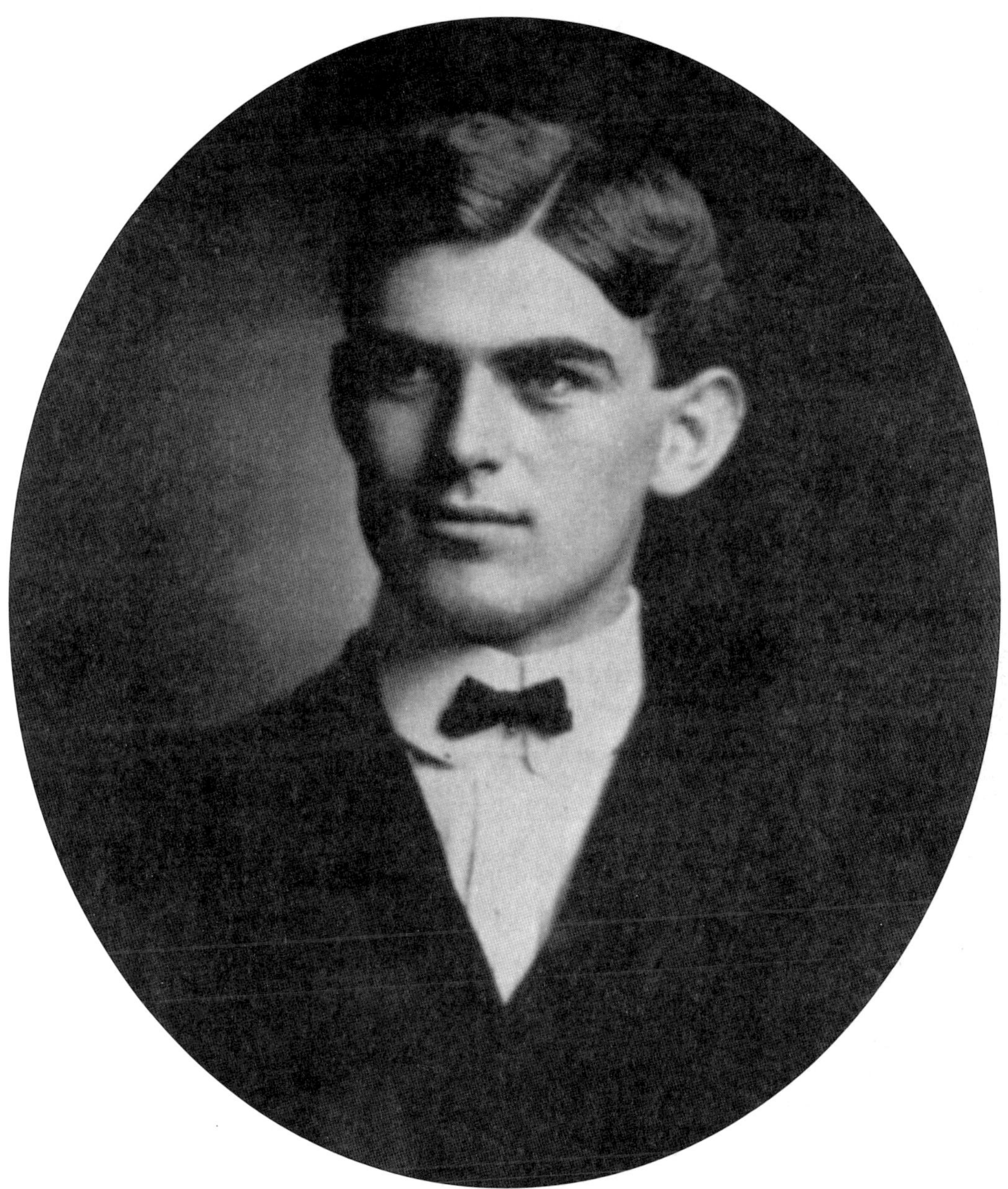

Bernhard C. (B. C.) Ziegler is seen as a young man around 1900. Born in Washington County in 1884, Ziegler began his first business selling insurance while still in high school. This business was the starting point for several other enterprises, including the founding of the West Bend Aluminum Company in 1911 at the age of 27. By 1914, Ziegler was devoting so much time to the West Bend Aluminum Company he took over management and in 1921 was elected president. He served as president and director until his death in 1946. He became one of the most influential people in West Bend, serving as president of the First National Bank of West Bend, chairman of Gehl Brothers Manufacturing Company, president of the West Bend Mutual Fire Insurance Company, and director for the Wisconsin Manufactures Association. Throughout his remarkable life, Ziegler was described as a man of vision with the ability to see those visions through to fruition. He was forceful, methodical, and resourceful, a true entrepreneur.

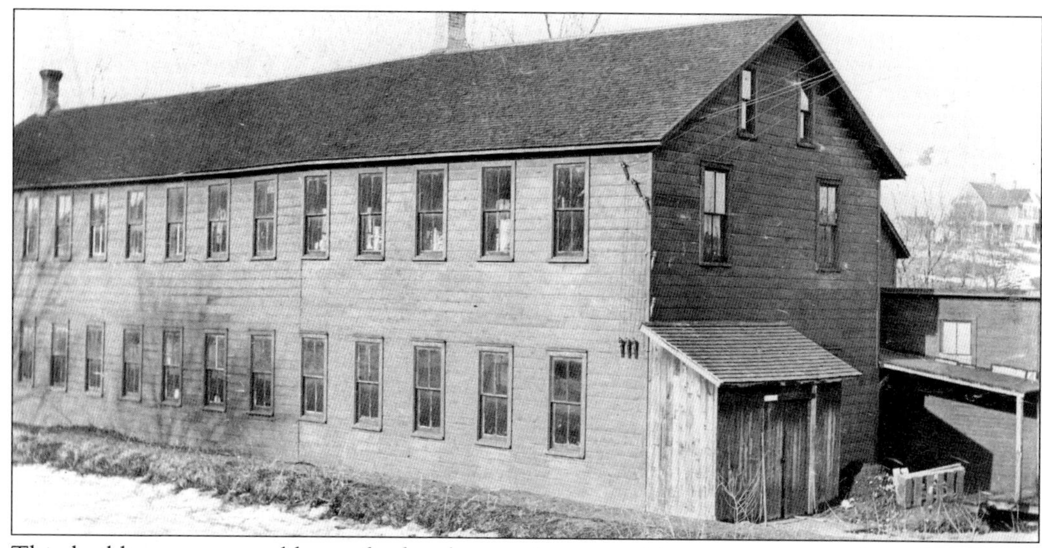

This building was a workhorse for local manufacturing. The original site of the West Bend Aluminum Company from 1911 to 1914, the building was the starting place of many West Bend manufacturers. Beginning in 1897, A. Wostal and Company manufactured pearl buttons at this location. Over the years, the West Bend Knitting Company, Enger-Kress Pocketbook Company, West Bend Barn Equipment Company, Carl Pick Manufacturing Company, and West Bend Plating Works all inhabited the site. West Bend Plating Works provided plating services to the West Bend Company for many years. The building was abandoned in 1989 and was demolished in 1997.

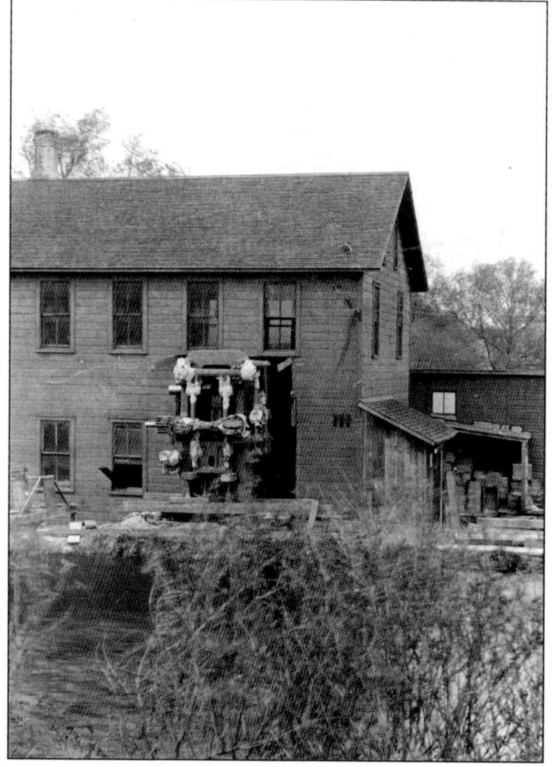

This photograph shows the above building when it was the site of the first West Bend Aluminum Company factory. Here the first large press is moved into the first factory site. A large hole had to be cut into the wall of the building to facilitate the move. This press weighed about 10,000 pounds and could turn out 15 different types of utensils.

Carl Wentorf is standing on the Bullard boring mill, one of the first large pieces of machinery to be moved into the earliest West Bend Aluminum Company factory site. Originally tool and die makers for the Aluminum Goods Manufacturing Company in Manitowoc, Carl and Robert Wentorf moved to the West Bend Aluminum Company after being laid off.

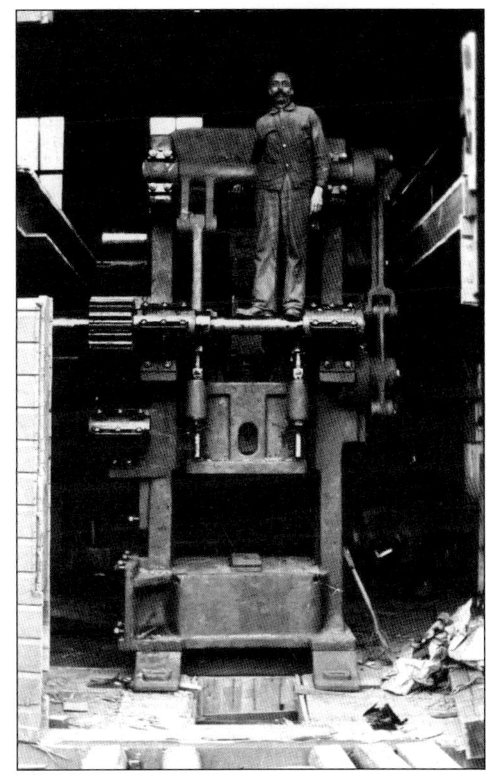

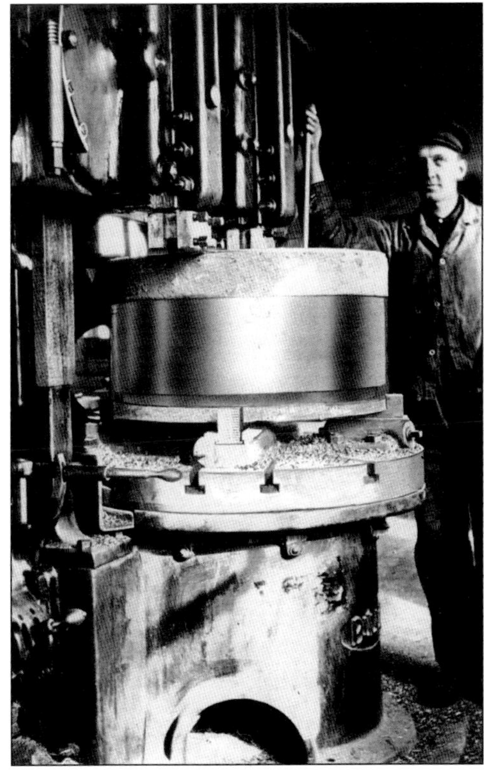

The Bullard boring mill was the largest machine in the first West Bend Aluminum Company Factory. Here tool and die maker Robert Wentorf poses with the mill.

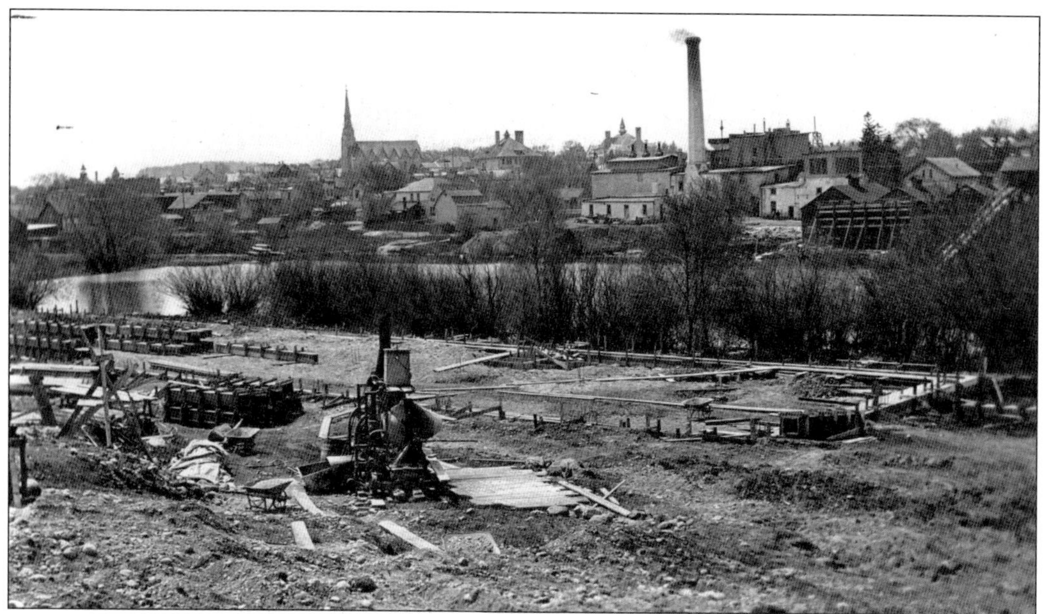

By 1914, the success of the previous three years allowed the West Bend Aluminum Company to begin construction on a new factory. A hill on the east side of the Milwaukee River was chosen for the new site. The hill was graded down and fill was used along the riverbank to stabilize and extend the area. Here the foundation of the factory is under construction.

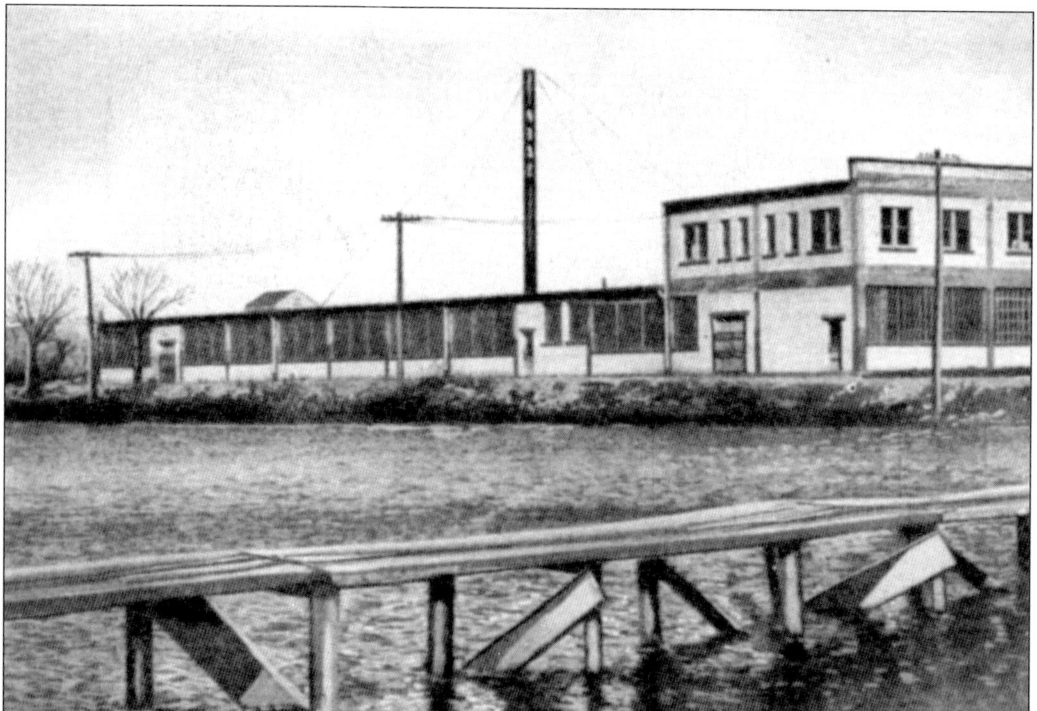

A temporary footbridge crossed the Milwaukee River around 1915. The bridge was used by workers to reach the newly constructed West Bend Aluminum Company factory until a more permanent route was constructed.

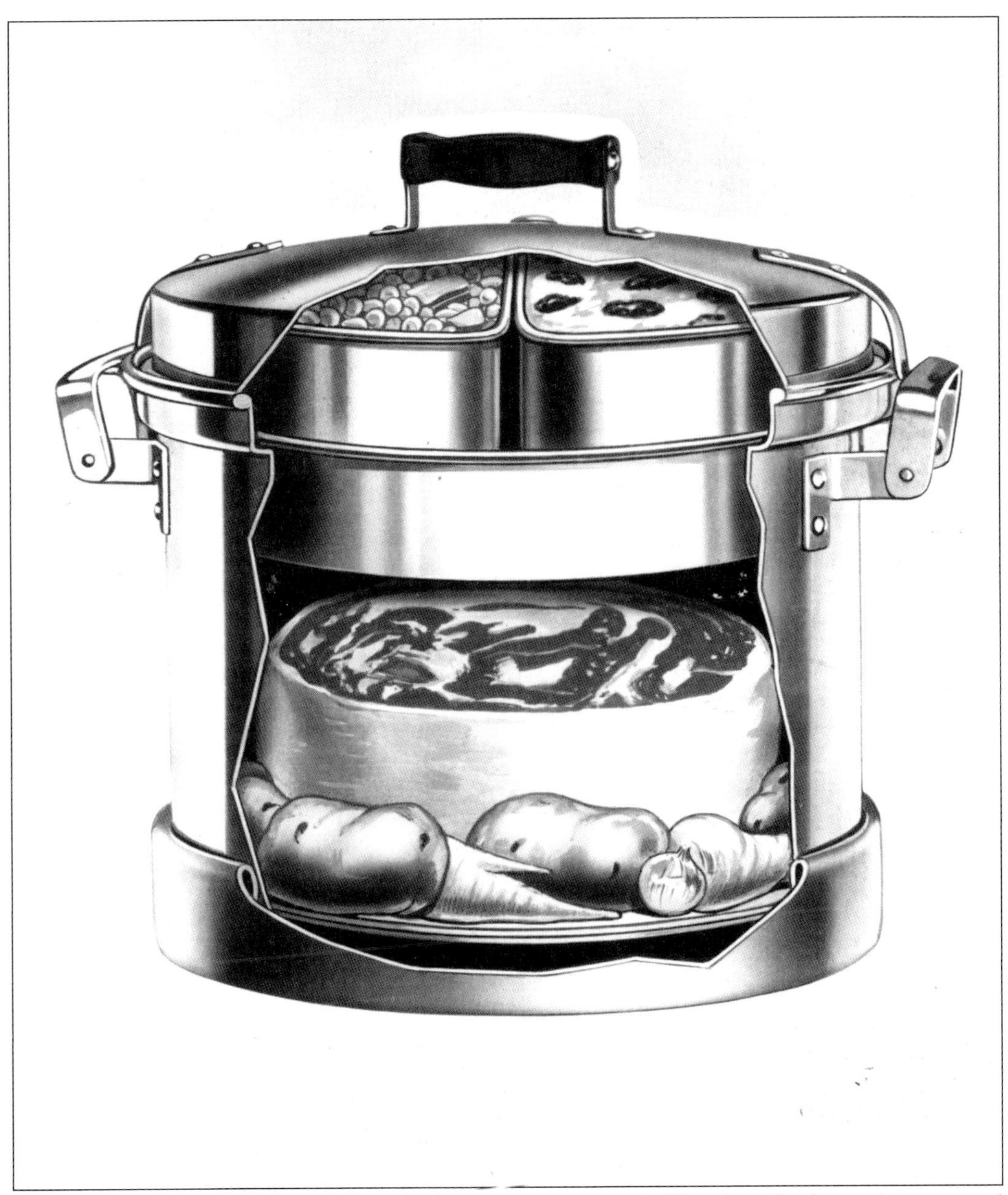

This cut away view of the West Bend Aluminum Company's Waterless Cooker is from around 1922. One of the company's best sellers, this view shows the convenience of preparing a complete meal in one pot. The first Waterless Cookers were stamped from heavyweight aluminum and sold with a cast-iron base for cooking on wood or coal stoves. The success behind the Waterless Cooker was in its design. The lid, which was fitted with locking clamps, trapped steam during cooking and allowed food to be cooked in its own juices. The Waterless Cooker was advertised as a time and money saver. Advertisements depicted women leaving the home while their Waterless Cooker prepared a full meal within hours. The product was sold door to door through commissioned West Bend Company salesmen, and by 1941, sales reached over six million. The Waterless Cooker's popularity led to the development of a complete line of waterless cookware called Flavo-Seal.

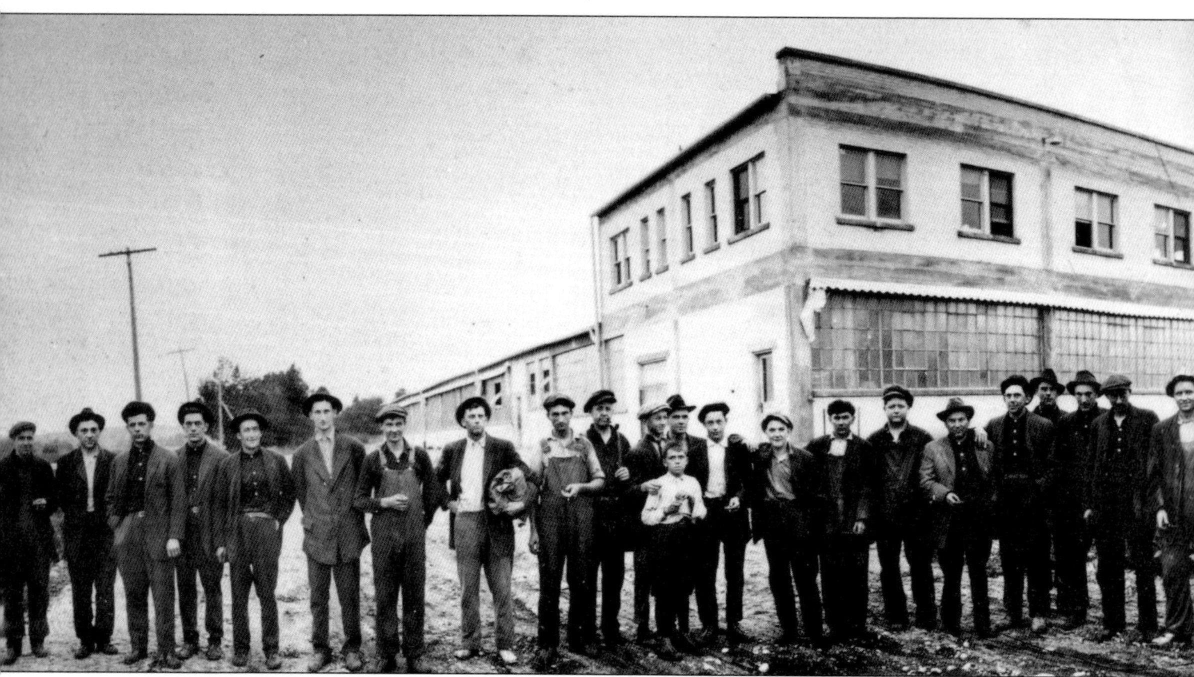

The first West Bend Aluminum Company staff to work in the new factory is photographed in 1914. The workers are, from left to right, Art Manthie, Joe Opgenorth, Al Hron, Frank Hron, Val Gonring, William Hackbarth, Alfred Bohn, William Fulwiler, Phil Carry, John Hartman, Ken Wiess, Elmer Wentorf, Oscar Schubert, Arthur Berger, Arnold Goetter, Frank Bassill, Carl Hirschboeck, Herman Winkler, Art Habeck, Waltern Manthei, Otto Oelke, John Kugler,

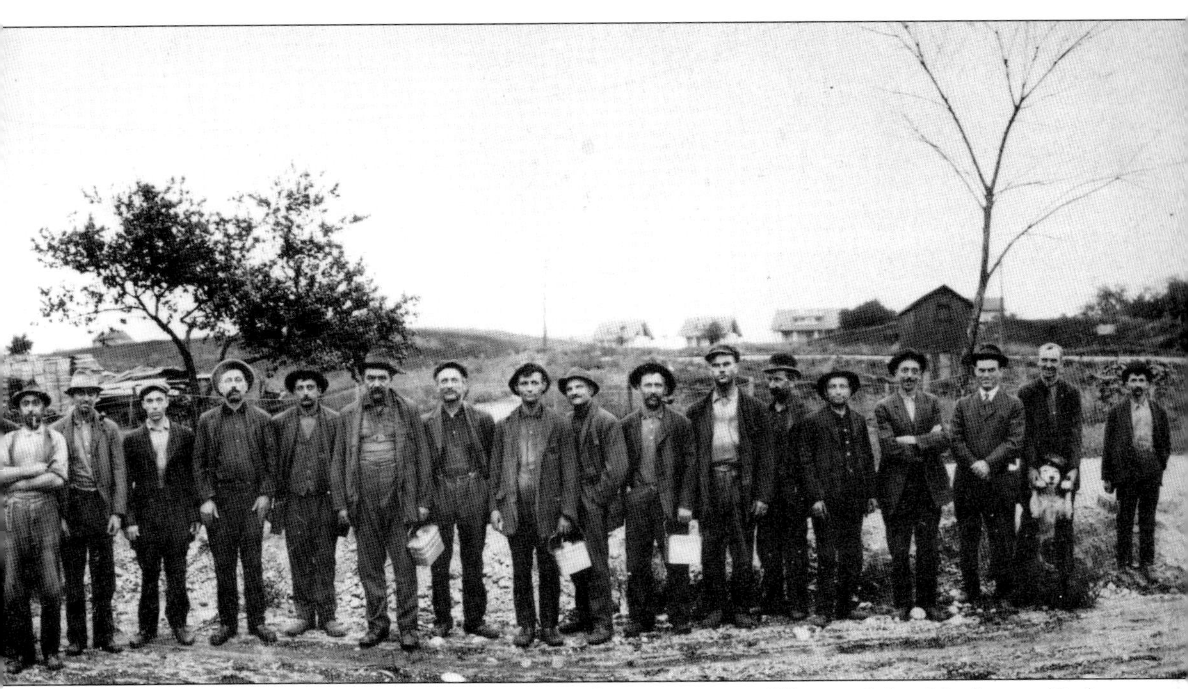

Helmuth Seeger, Robert Schnoeck, Louis Spanheimer, Martin Wagner, John Neubauer, Carl Weinkauf, Peter Mueller, Jacob Wagner, Herman Frank, Henry Baerber, Fred Schields, John Bahr, Adam Biertzer, Louis Kuehn, Fred Ehnert, Carl Wentorf, Bernhard C. Ziegler, Robert Wentorf, and Joe Karius. Several of these individuals stayed with the company for many years, some working their way up to management roles.

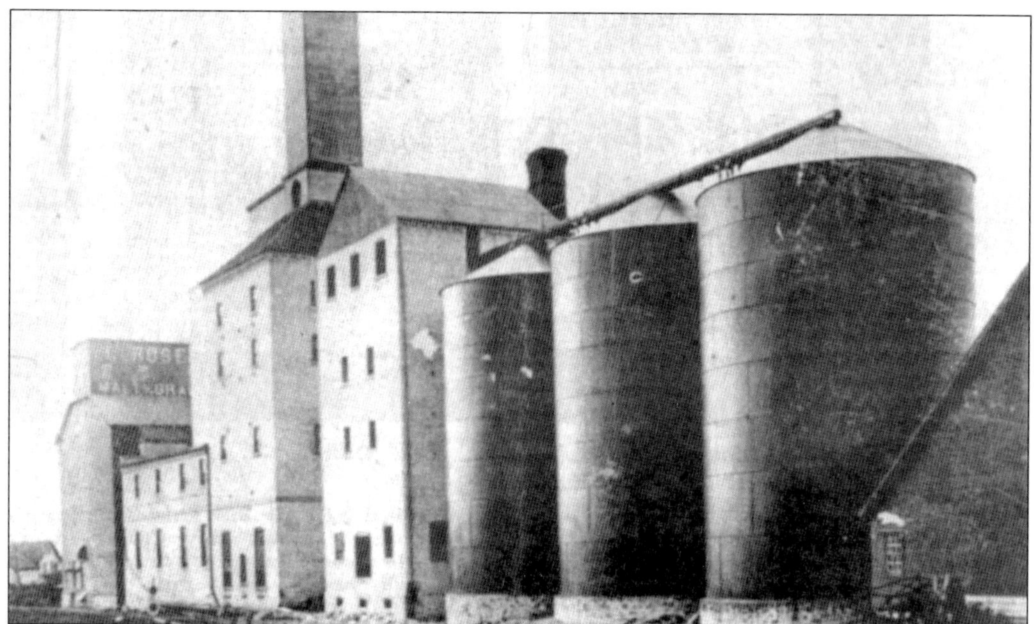

The L. Rosenhiemer Malt and Grain company began business in 1874. Like many malt companies, Rosenhiemer's fell on hard times when the 18th Amendment, establishing Prohibition, went into effect in December 1920. Taking a cue from his business rivals in West Bend, owner Adolph Rosenhiemer looked toward a new business in the aluminum industry to supplement his income. The Kewaskum Aluminum Company was born.

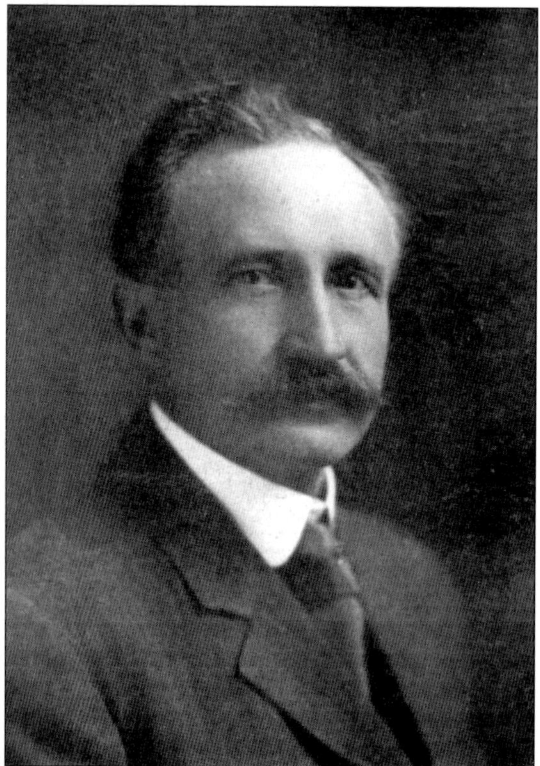

This portrait of Adolph Rosenhiemer was taken around 1930. He was born in 1861 in Slinger. After moving to Kewaskum to help his brother run the general store, he became active in local politics. He served as the village of Kewaskum's first president as well as on the Washington County Board of Supervisors. At the time of his death in 1942, Rosenhiemer had been instrumental in starting several local businesses and held positions on several local boards.

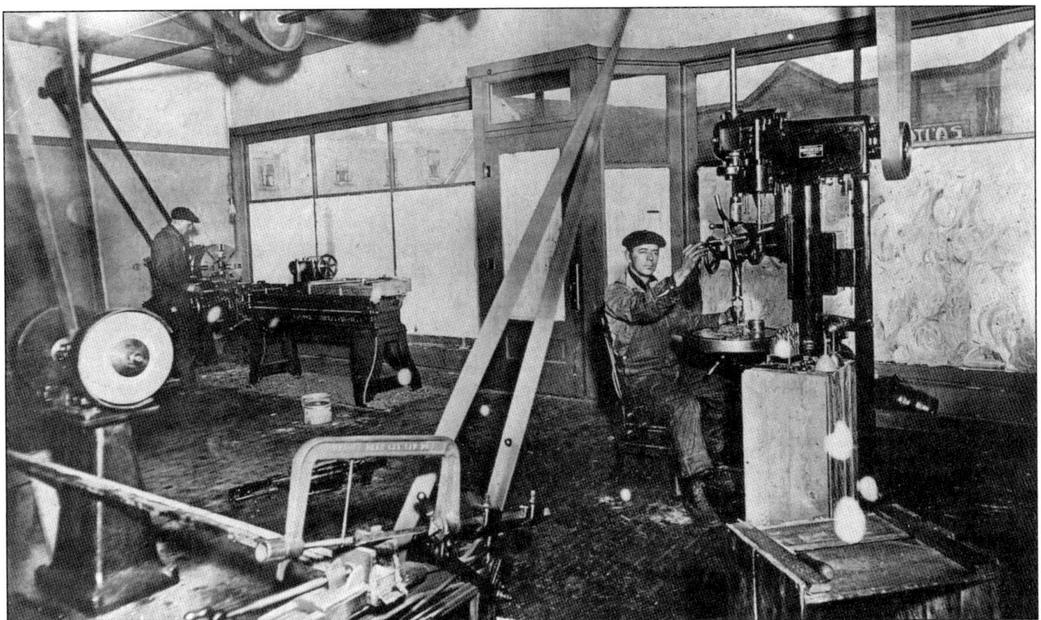

The interior of the first Kewaskum Aluminum Company factory, this photograph was taken in 1919. Working on tooling before the new factory is constructed, Al Hron, seated, and Art Manthei made do in temporary quarters. Located in a garage on Fond du Lac Avenue in Kewaskum, the building was used as the advertising department for Kewaskum Aluminum Company and later Kewaskum Utensil Company and Regal Ware Inc.

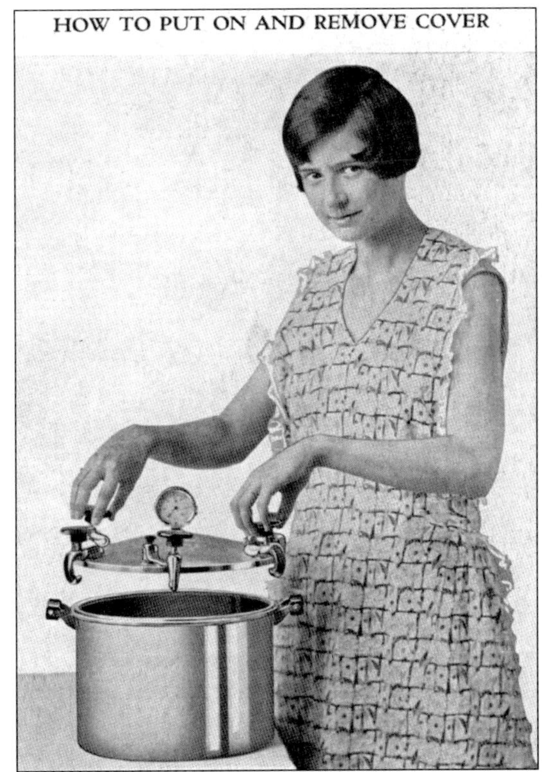

The Liberty Steam Pressure Cooker was manufactured by the Kewaskum Aluminum Company during the 1920s. The pressure cooker was advertised as being an economical whole meal cooker and as the most versatile utensil in the kitchen. It was also advertised as a convenience as the cook could leave the house while food cooked because it would not burn.

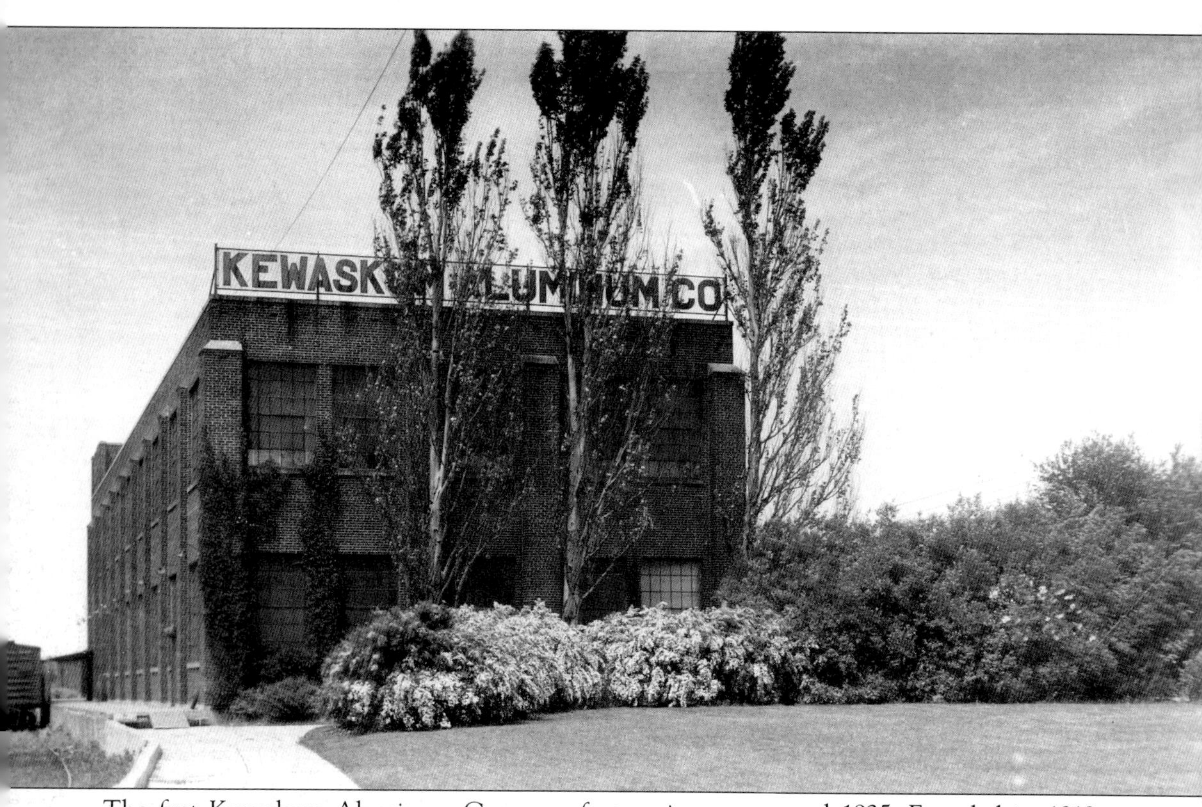

The first Kewaskum Aluminum Company factory is seen around 1925. Founded in 1919 in response to Prohibition, work immediately began on construction of a new factory on Altenhofen Street, now First Avenue. Production began in August 1920 with 60 employees. After only 18 months of operation, Kewaskum Aluminum Company earned the Good Housekeeping Seal. The manufacturer saw rapid growth in the first years. In 1925, a fire caused a minor setback but as the building was fireproof, little structural damage was done; however, inventory damage was heavy. The company bounced back only to be challenged again with the Great Depression. Kewaskum Aluminum Company tried to keep employees busy during the economic downturn. Grounds maintenance and painting the building were two of the ways jobs were maintained during times of low production.

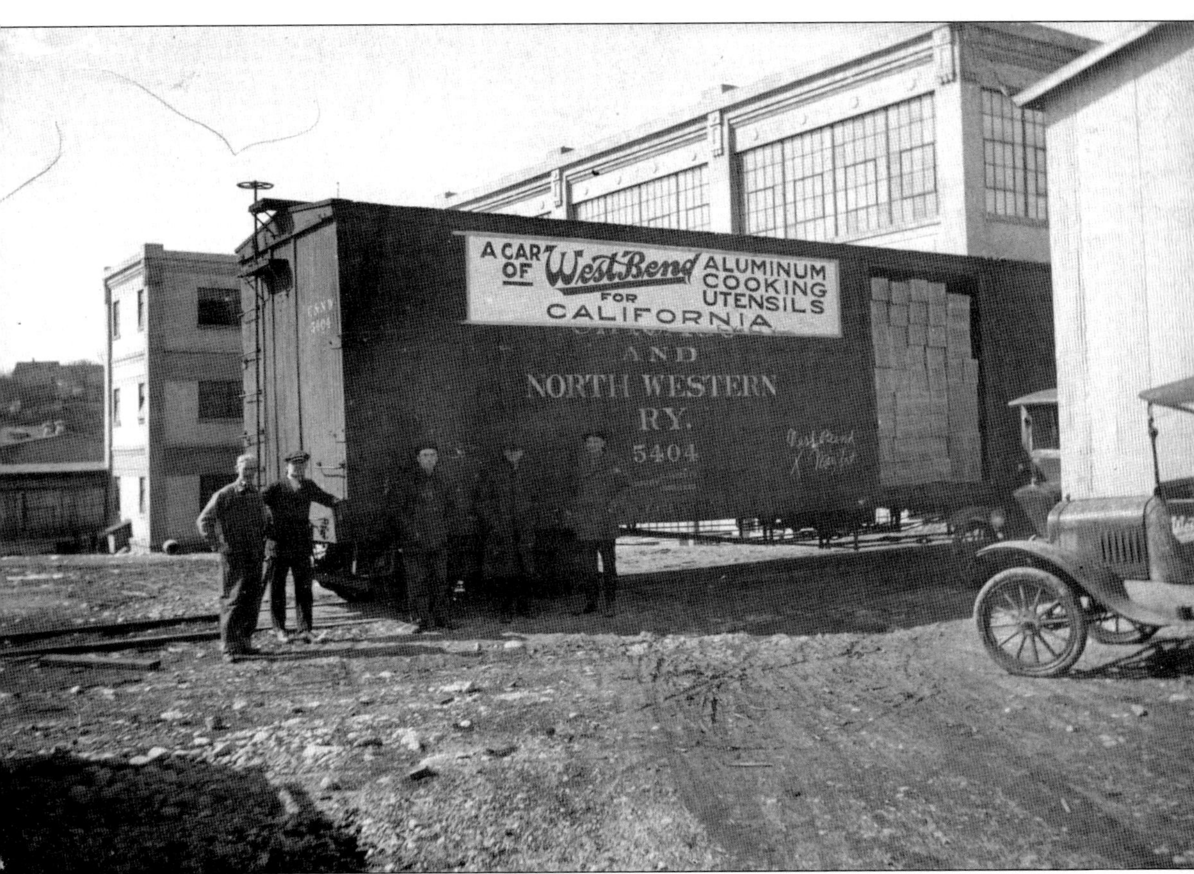

This was the first train car load of West Bend Aluminum Company product ready for shipment to California. The shipment caused quite a stir as it was a symbol of progress and success for the company. This was only the first of many cross-country and eventually worldwide shipments.

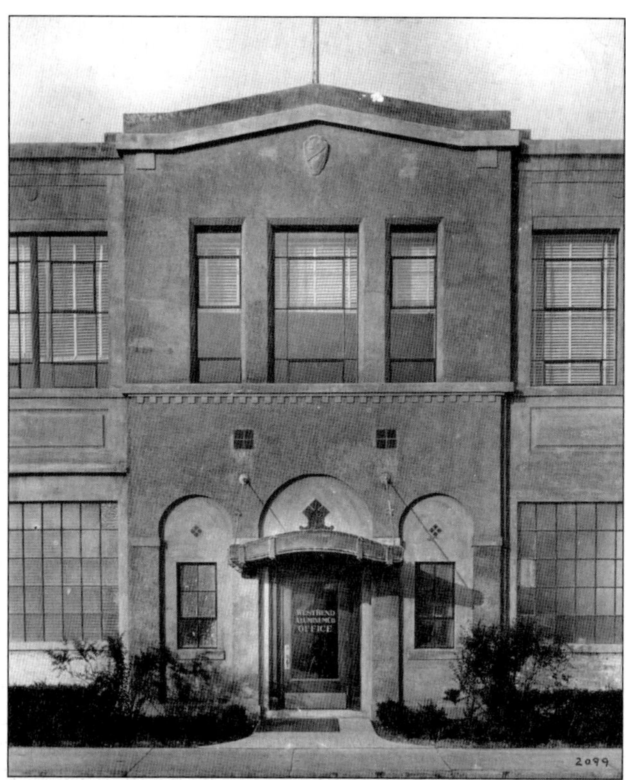

The West Bend Aluminum Company's newly completed office building is seen in 1927. A three-story section began construction in 1926. The addition added manufacturing space as well as a new office area. It also created enough jobs to elevate the West Bend Aluminum Company to be the largest employer in West Bend.

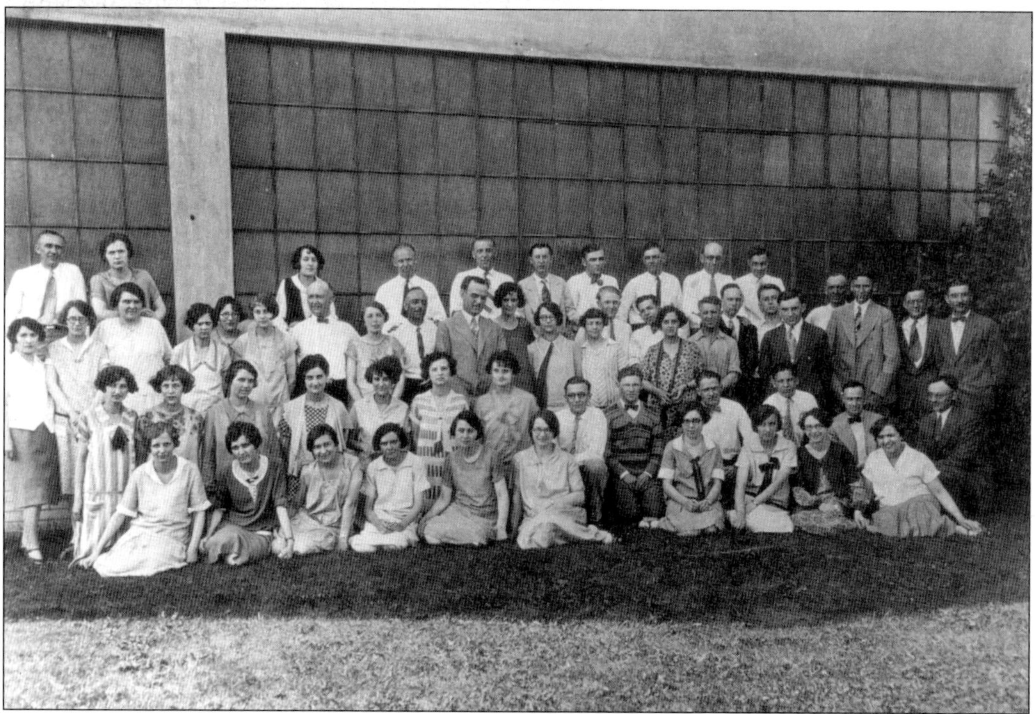

The West Bend Aluminum Company office staff of 1927 poses outside the new quarters. The new exterior was designed to provide a professional welcome for the business side of the company.

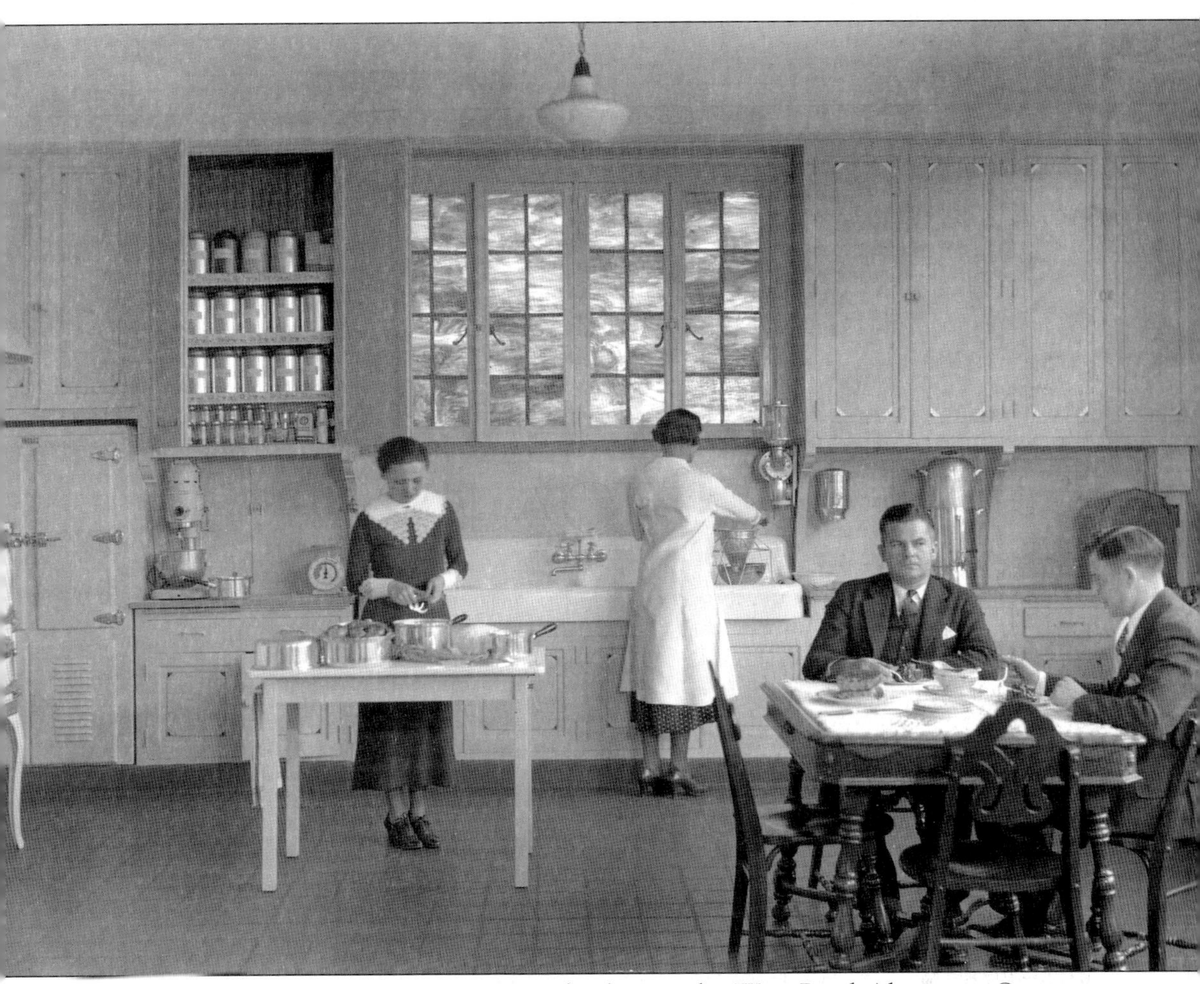

This photograph shows the innovative test kitchen at the West Bend Aluminum Company. In the 1920s, the company installed a test kitchen to experiment with products in real life conditions. This was also one of the first locations, along with the office, that the company employed women from an early date. The title of home service director went to the woman in charge of running the test kitchen. The department was in charge of keeping contacts with women's page editors, home economic teachers, women's magazines, and giving presentations to high school and college students. It was here that salesmen were trained to use the products and to practice techniques to improve their sales. The Home Service Department also prepared special luncheons for plant guests and compiled materials for recipe booklets.

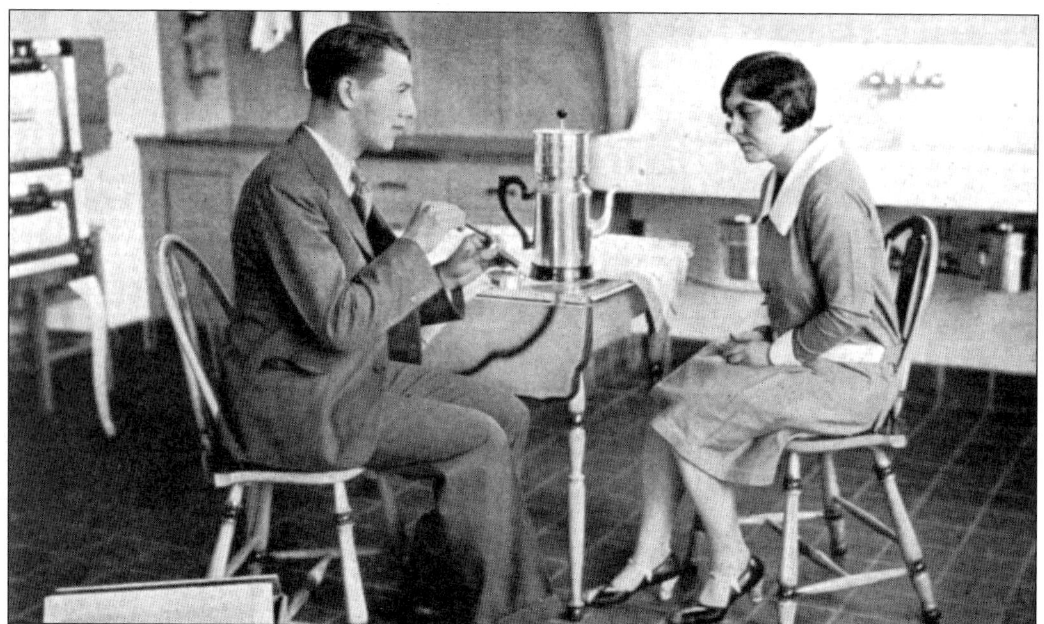

A salesman in the 1920s talks to a housewife to try and sell the Deluxe Arrow Coffee Maker. This photograph comes from a booklet given to salesmen. The booklet contained step-by-step instructions on how to sell the coffeemaker. It included answers to any objections encountered during a sale. The door-to-door, or direct-sales method, was the way many of the early West Bend Aluminum Company products were distributed.

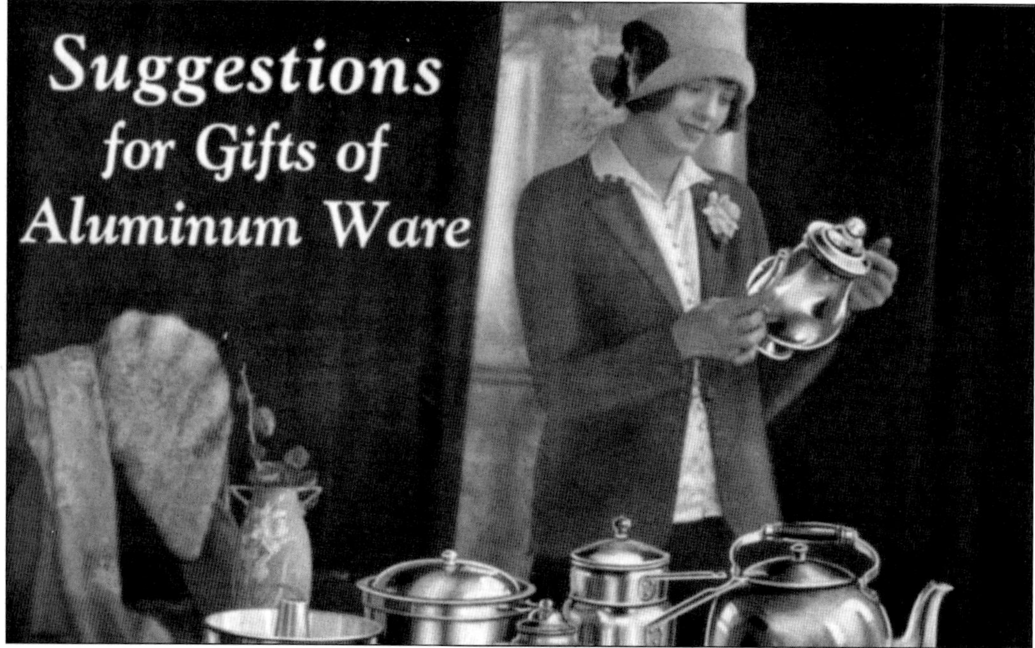

Seen here is an advertisement for giftware printed in the West Bend Aluminum Company's catalog from 1926. As competition grew, companies had to become creative in their advertising and design. An increase in choice saw a shift in consumer demands. Items such as cigar cases and serving trays are examples of giftware items marketed to the higher-end consumers.

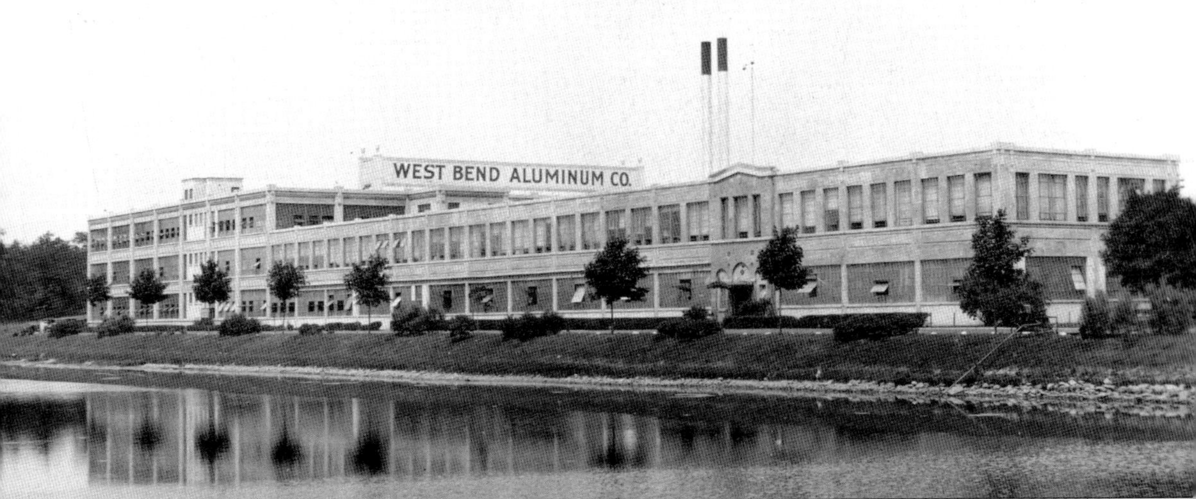

This photograph shows the completed factory and office building additions of the West Bend Aluminum Company in 1927. During the first three decades of existence, the West Bend Aluminum Company expanded three times. In 1918, the manufacturing space was increased with a three-story addition to the north end of the building. In 1927, a second three-story manufacturing space and new office building were added. Finally in 1937, a third three-story addition was constructed and the hallmark tower was added.

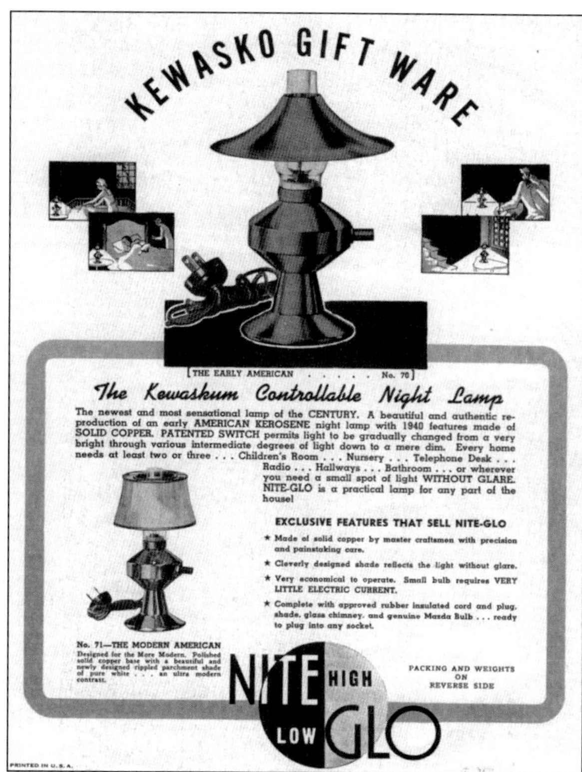

This Kewasko giftware advertisement from the Kewaskum Aluminum Company was used in a trade magazine. Many of the early products made from aluminum were novelty items. Combs, key fobs, lamps, and cigar cases were popular aluminum items.

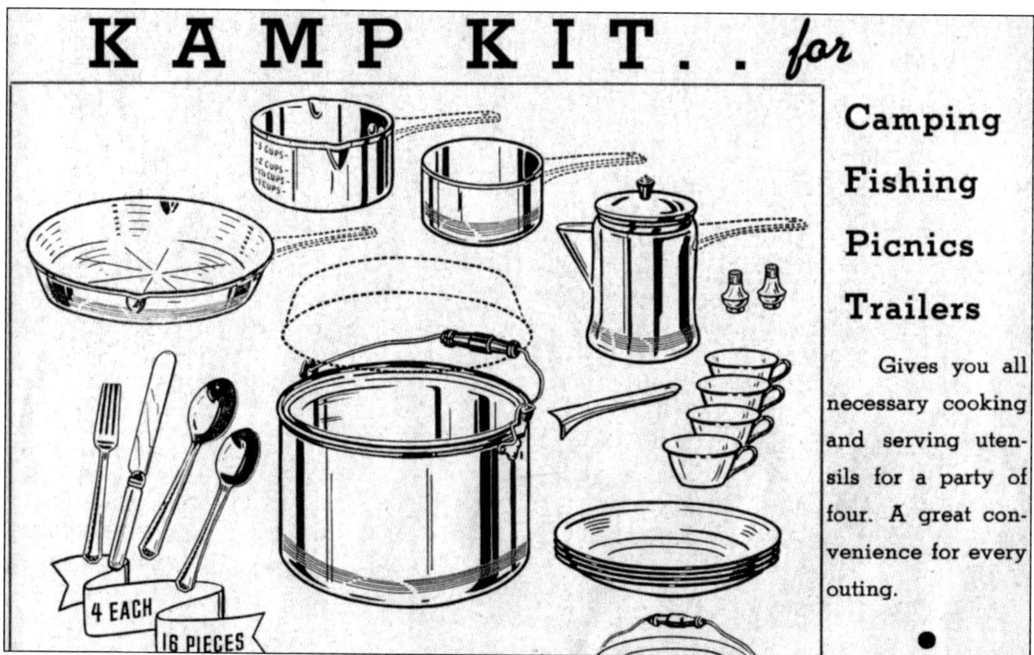

This advertisement from around 1940 marketed the Kewaskum Aluminum Company's Kamp Kit. The kit boasted 25 pieces that could all be packed into one kettle for easy transport. An interchangeable handle allowed for the use of several different utensils but broke down into a small space. Serving utensils for four and salt and pepper shakers were included.

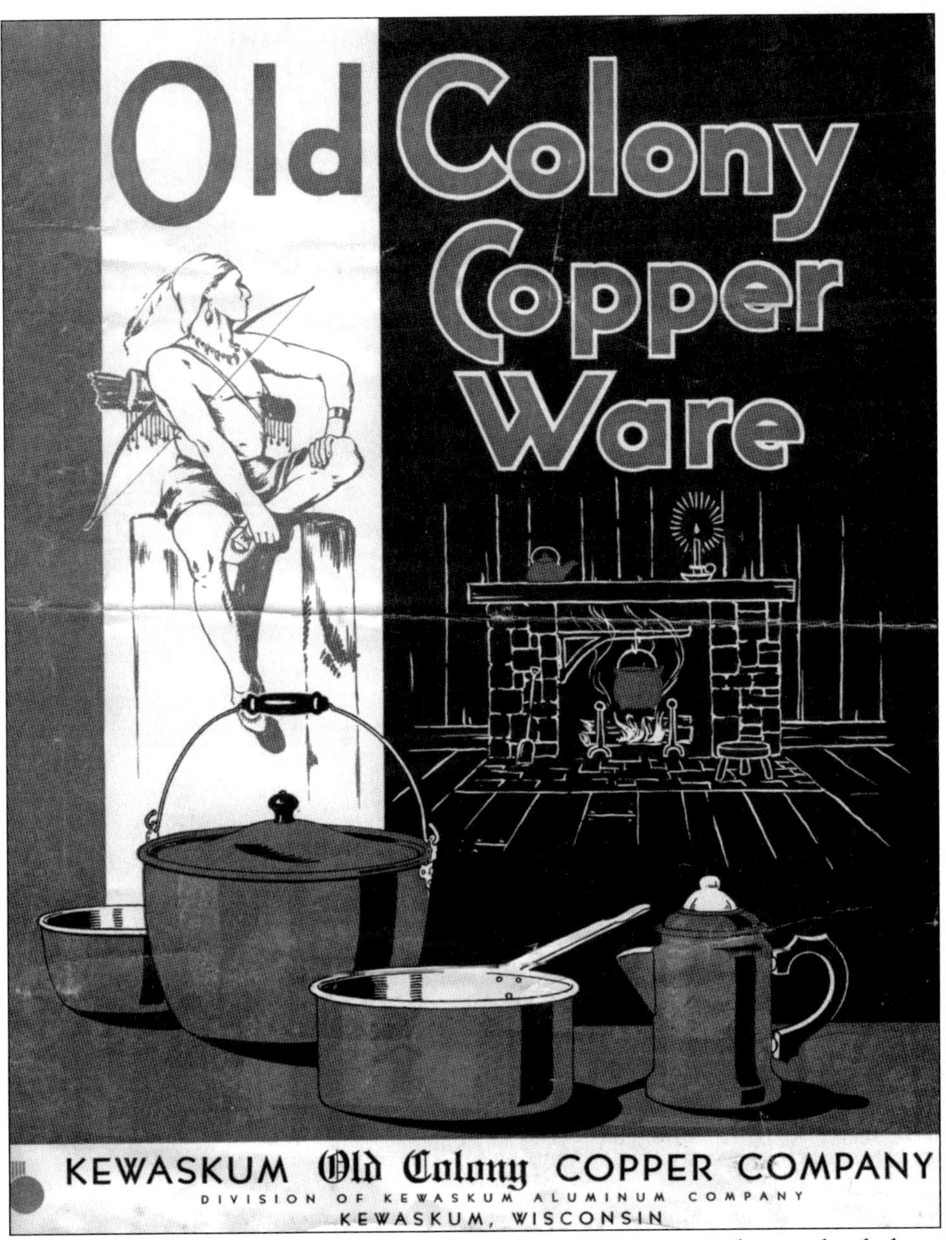

This Old Colony Copper Ware catalog cover from about 1930 is a typical example of advertising from the time. During the Great Depression, many aluminum manufactures supplemented their goods with copper. Copper products became a status symbol and luxury goods of the upper income households during times of economic hardship. In part due to the economic crisis, the Kewaskum Aluminum Company saw no profit from 1929 to 1939. In an effort to turn the company around, the Kewaskum Old Colony Copper Company was created as a division of Kewaskum Aluminum in 1932. Although demand declined at the end of the 1930s, in 1933 the copper ware comprised over 66 percent of the year's sales. In a further effort to rebound, the company opened another division. Stainless Steel Incorporated, the second Kewaskum division, began producing stainless steel cookware in 1938. By 1941, the company was well on its way to recovery depending mainly on the sale of stainless steel utensils.

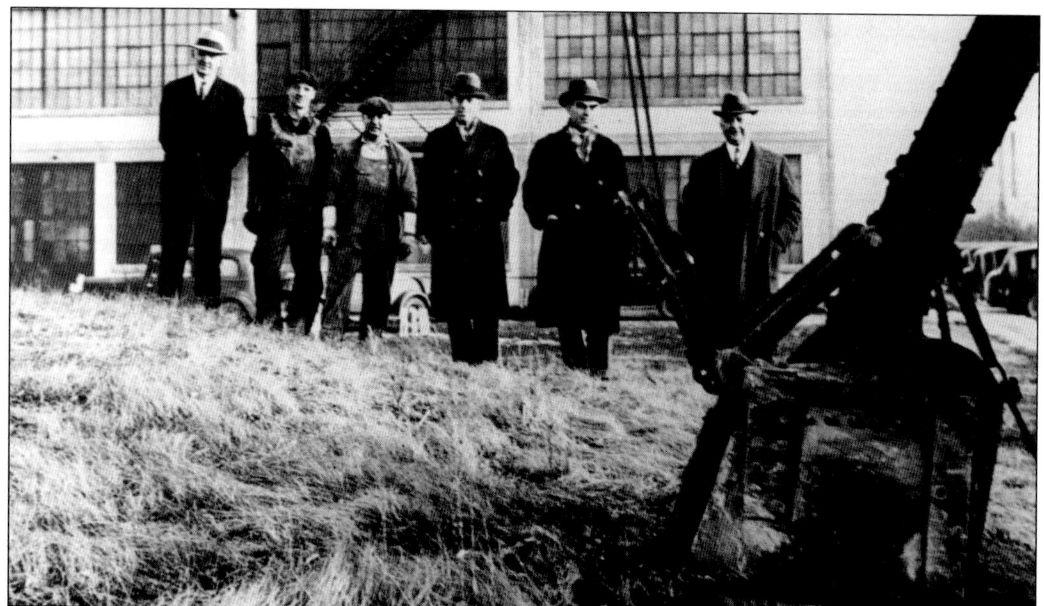

West Bend Aluminum Company officials look on as ground is broken for the 1937 factory addition. Although the United States as a whole was in a recession, the West Bend Aluminum Company continued to grow. The increased demand for the waterless cookware created a need for a three-story addition to the production facilities. It was during this addition that the trademark tower was added to the factory. The tower housed equipment that allowed for a larger exhaust system. This system removed dust and fumes from the manufacturing floor.

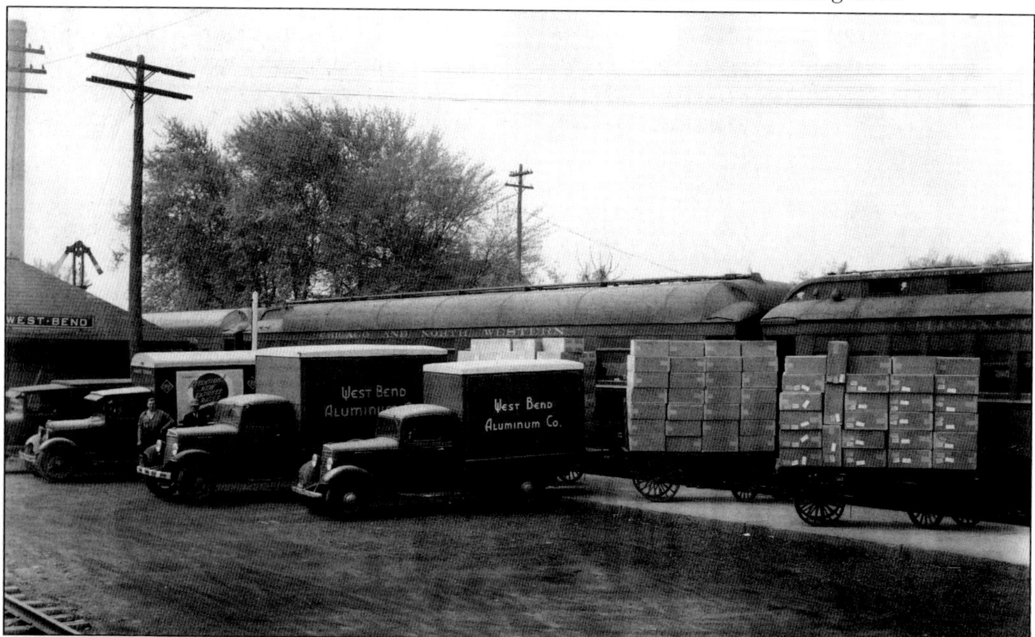

The West Bend Aluminum Company trucks load a train at the West Bend train depot around 1940. Although the company had a direct rail line past the plant, rush orders would be shipped via Railway Express Company from the depot. The Railway Express Company ran as part of the passenger train system.

Two

THE WAR YEARS 1935–1950

Even before the United States entered World War II, aluminum was moved to the mandatory priority list, reserving all of the material for war production. In Washington County and across the nation, manufacturers were rethinking their production methods. Some aluminum manufactures, including Kewaskum Aluminum Company, switched to stainless steel for a short time before it too was put on the reserve list. The lack of raw materials for domestic production forced many companies to move to government contracts to keep working. The major obstacle with government contracts was the task of conversion. Extensive retooling was required. The manufacture of wartime materials was much more precise than many companies were used to. One small error in production of a cartridge case could jam the weapon, and put troops in danger.

In June 1941, West Bend Aluminum Company secured a contract with the U.S. Navy to manufacture a 20-millimeter antiaircraft cartridge case. Conversion to wartime production began and soon around-the-clock, seven-day-a-week shifts were employed on the government contract.

Kewaskum Aluminum Company won a contract with the U.S. Department of War for experimental production of cartridge cases. In 1942, with brass in short supply, the company worked on the development of 75-millimeter steel cartridge cases to supplement the usual brass casings. Cold drawn steel cases had never been manufactured before so much of the early work was trial and error. The government expanded the program, awarding Kewaskum Aluminum compliments from the U.S. Department of War and further contracts. By the end of the war, there were concerns about the company continuing. Reconversion to peacetime production and the availability of aluminum for civilian production were major concerns. It was decided to sell the business while it would still turn a profit. Several companies put in bids, but J. O. Reigle and associates won, and the final transfer was made on February 6, 1945.

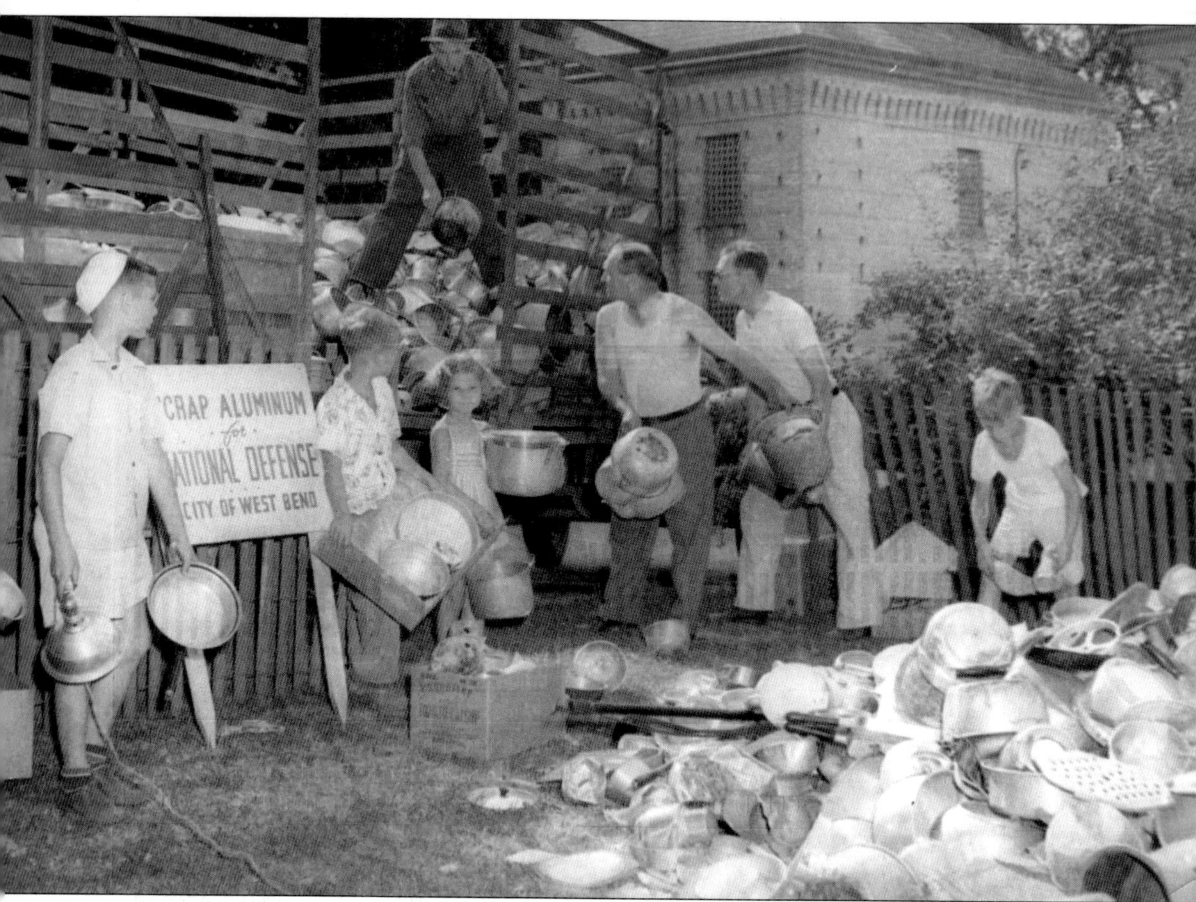

The city of West Bend's scrap aluminum collection site was located behind the Washington County Courthouse in July 1941. E. E. Skaliskey (center), chairman of the aluminum drive, loads discarded aluminum from the citizens of Washington County. Before and during World War II, many products were gathered in the name of national defense. The local newspaper announced that the aluminum would largely be used to manufacture aircraft. Under the Scrap for Victory campaign, other material collections included kitchen grease, newspaper, rags, rubber, and other metals.

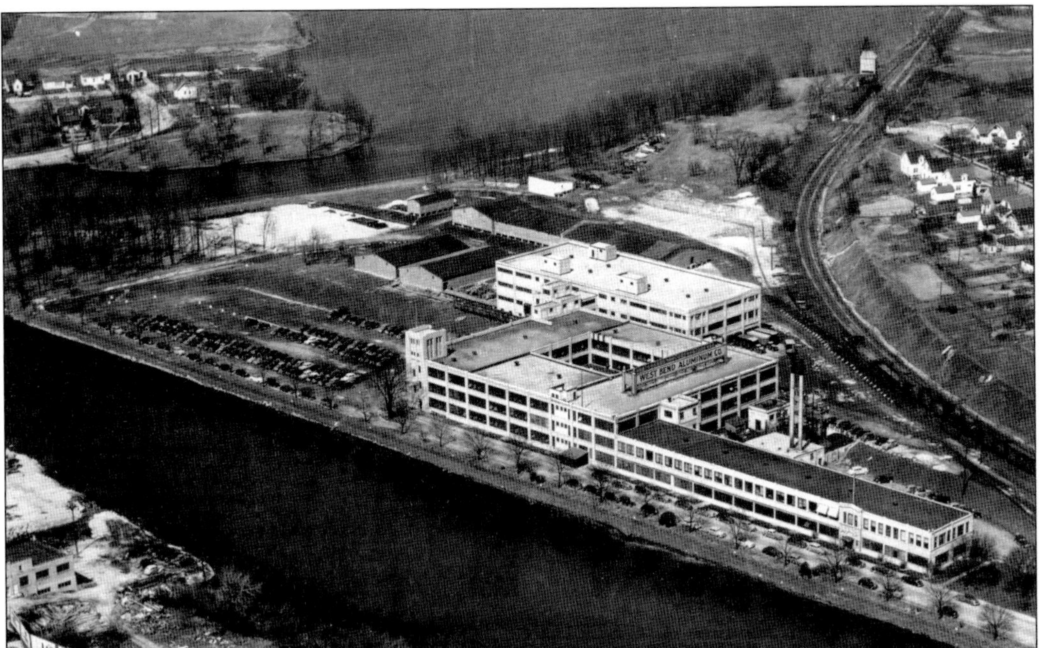

The aerial view of the West Bend Aluminum Company seen here was taken around 1946. The new office building has been completed and is the long addition to the right of the main building. Ground was broken that same year for a new four-story warehouse traffic building.

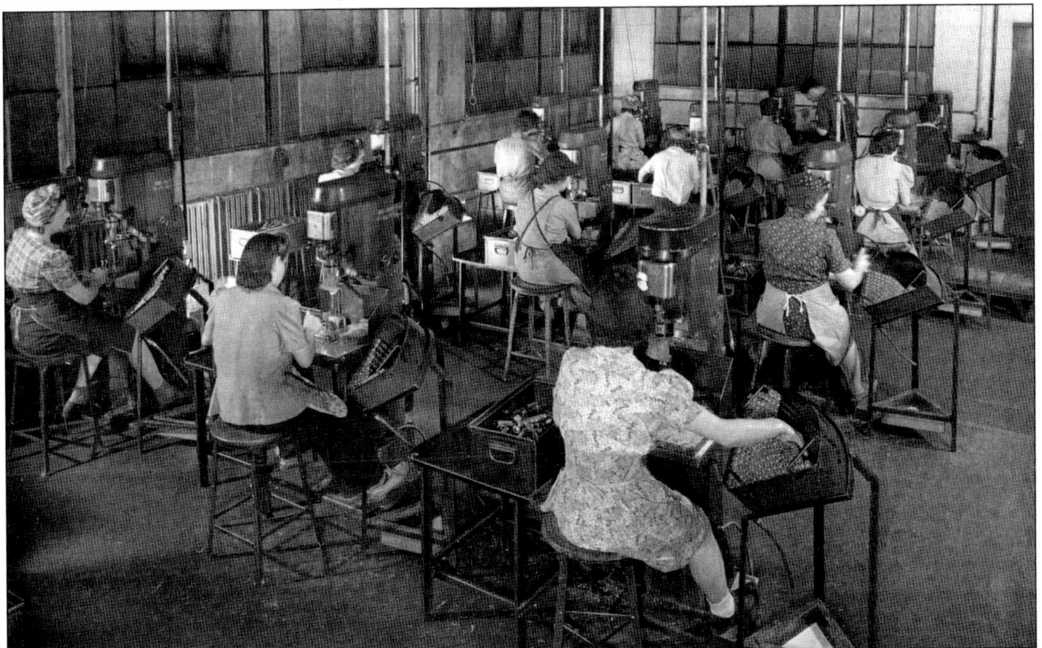

In June 1941, West Bend Aluminum Company secured a contract with the U.S. Navy to manufacture 20-millimeter Oerlikon antiaircraft cartridge casings. Conversion to wartime production began, and by December 1941, two shifts were employed under the government contract. This photograph shows women at the West Bend Aluminum Company around 1942 producing the 20-millimeter casings.

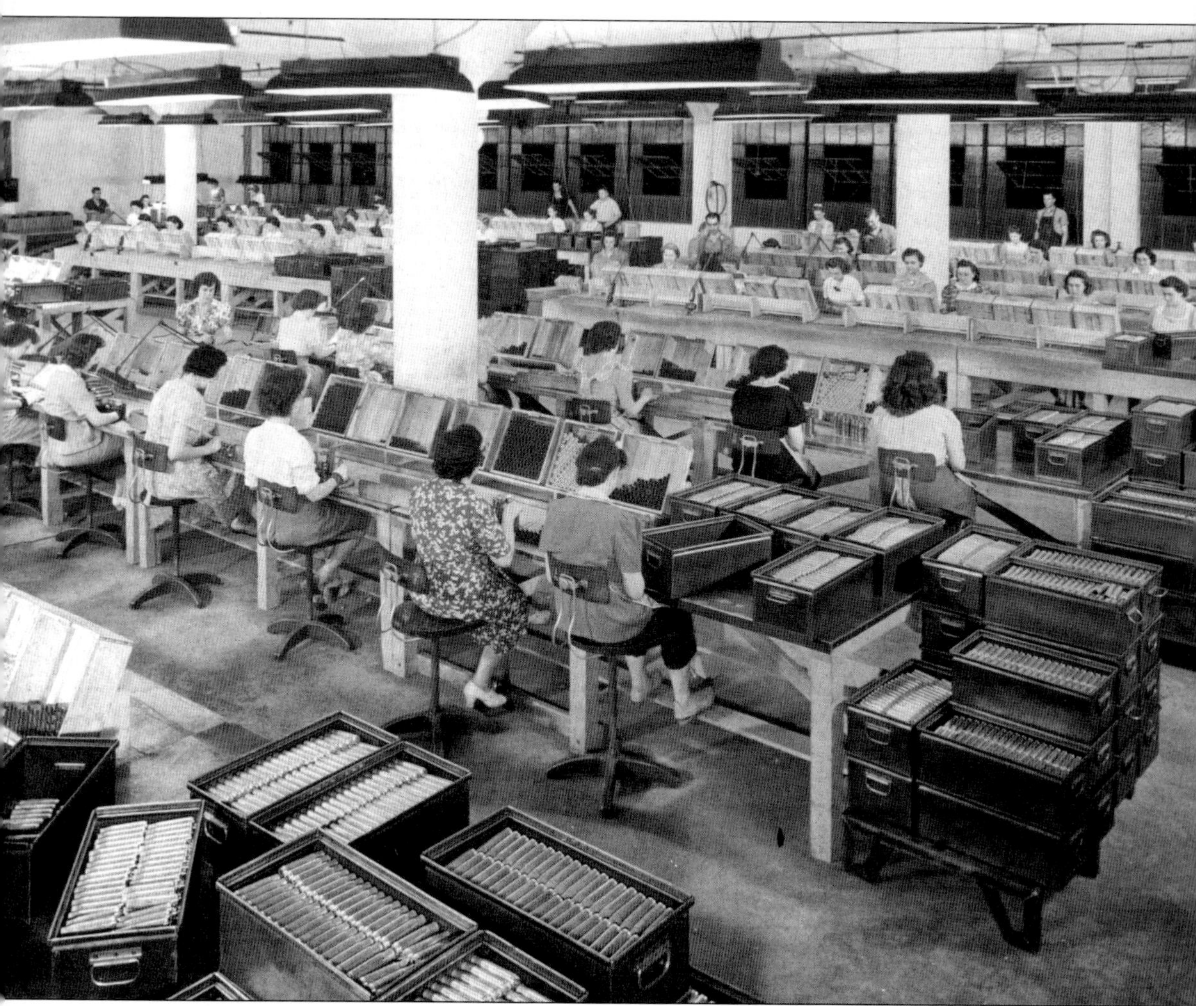

This photograph shows women working in the West Bend Aluminum Company around 1942. As World War II continued, it became necessary for women to go to work to replace the men who were on the front lines. In addition to running their homes, many women worked 48-hour weeks and overtime to help their loved ones win the war. The introduction of women into jobs traditionally held by men changed how society viewed the feminine role.

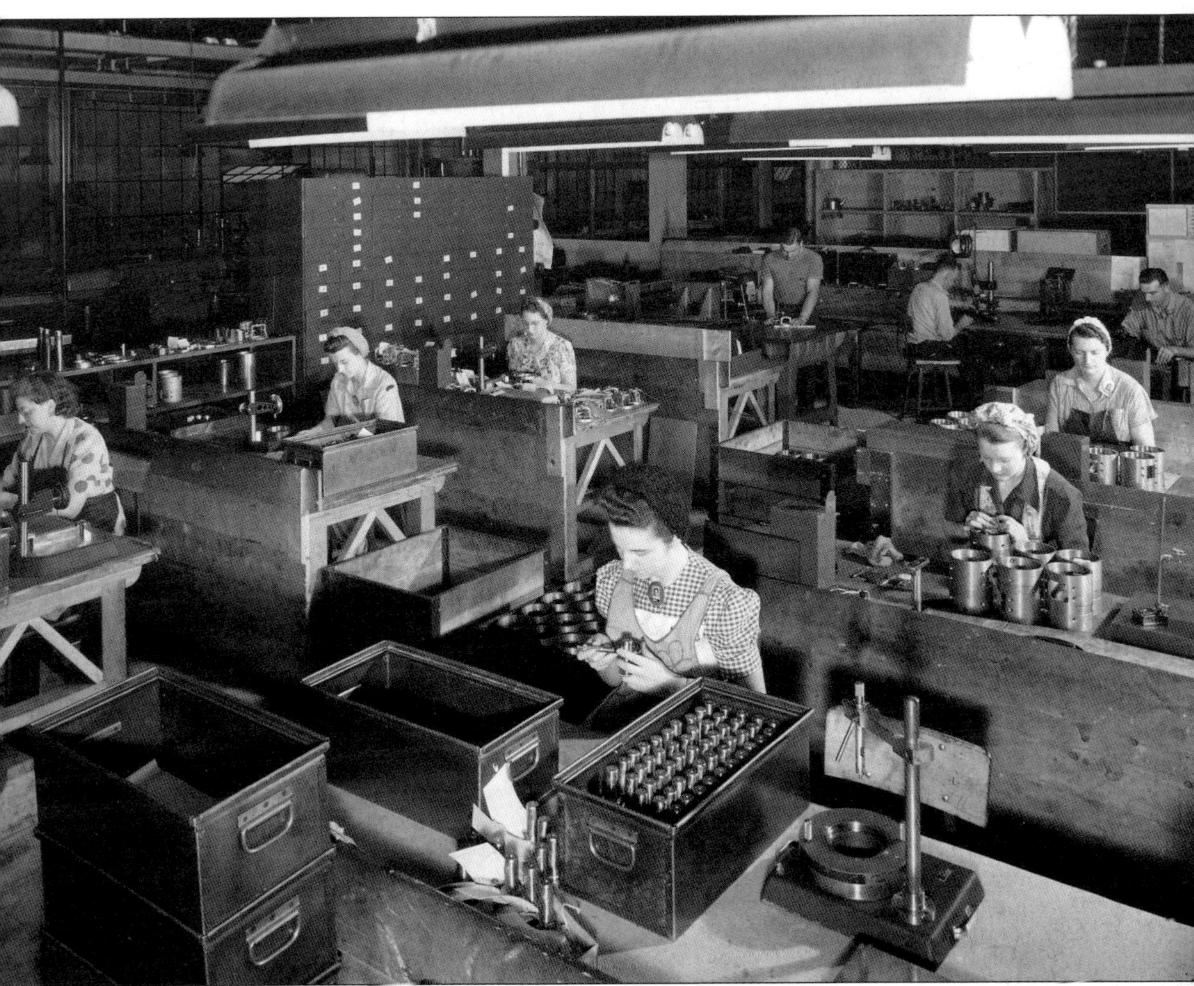

Many women worked in the West Bend Aluminum Company around 1942. This photograph shows wartime production of 20-millimeter Oerlikon antiaircraft cartridge cases. By this time, production ran 24 hours a day, seven days a week. With this schedule, the West Bend factory could produce over six million cartridge cases each month.

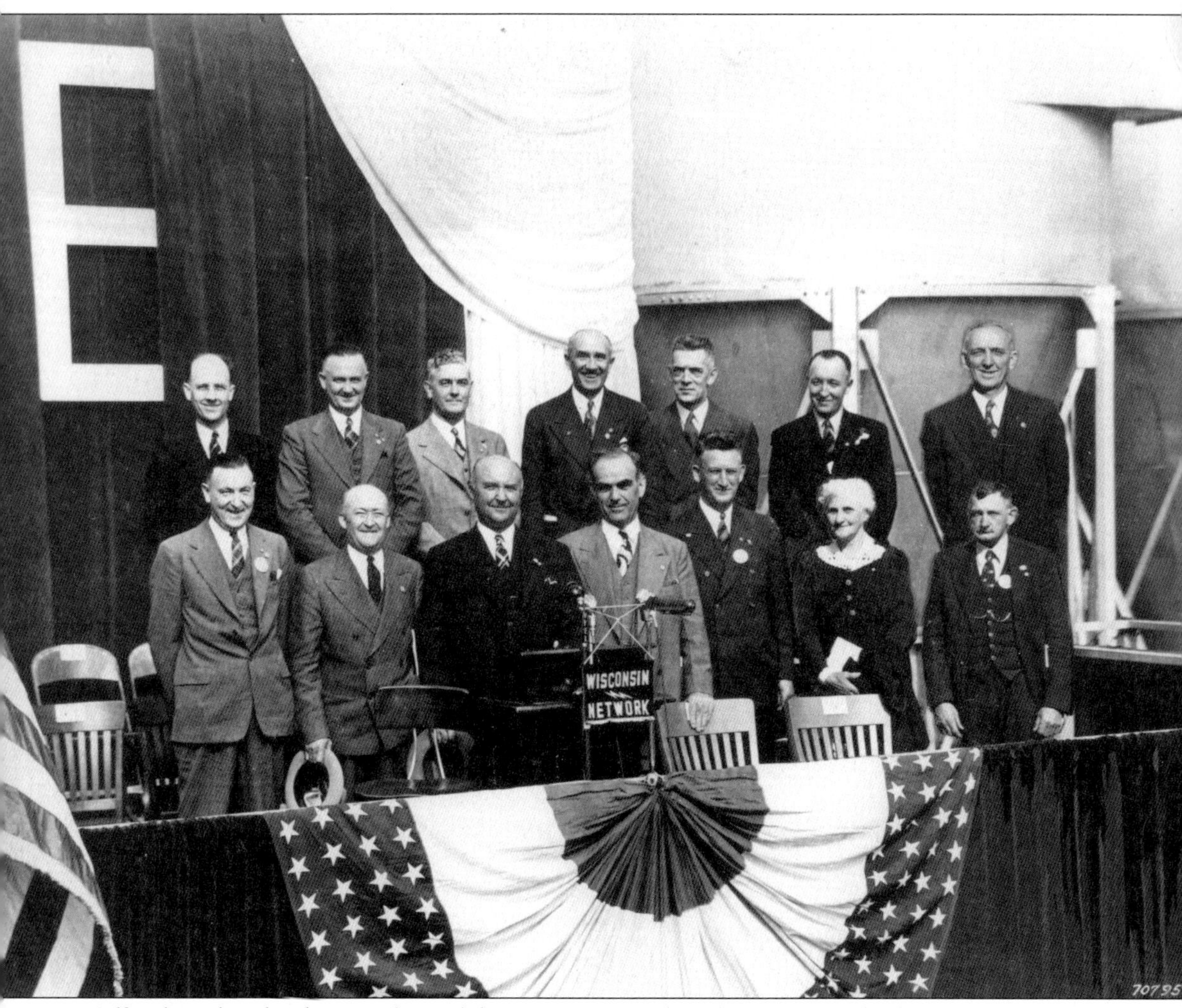

Officials gather for the awarding of the Navy E Award to the West Bend Aluminum Company in May 1942. Individuals from left to right are (first row) Otto Oelke, Val Gonring, A. G. Langenbach, B. C. Ziegler, William Hackbarth, Pauline Rusch, and Peter Mueller. The second row includes Ray Fellenz, Al Bohn, John Schmidt, A. C. Kieckhafer, and Frank Hron.

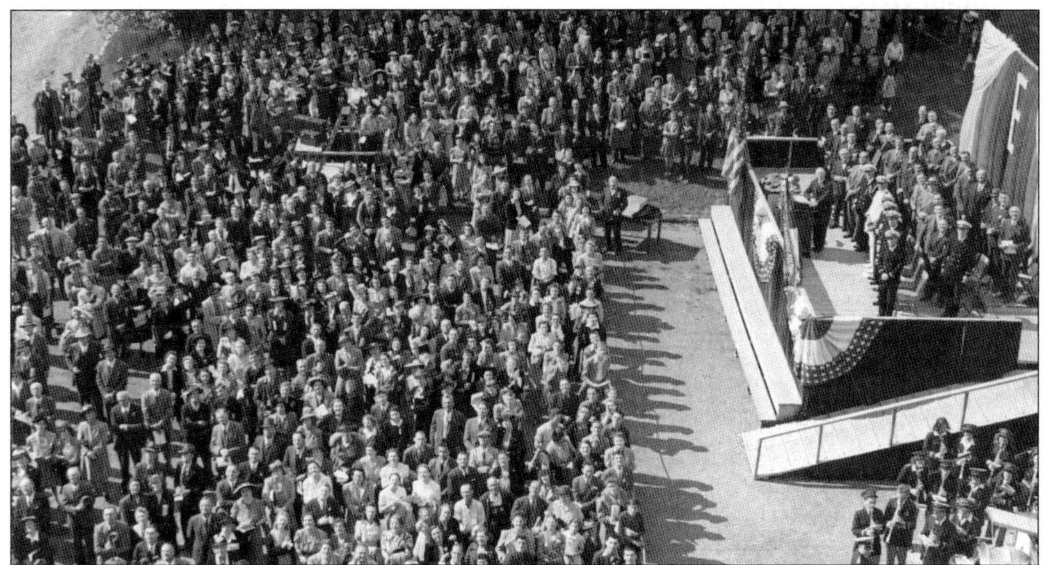
The Navy E Award was given by the U.S. Navy to businesses that showed excellence in production. In May 1942, over 2,000 people gathered to see this award presented to the West Bend Aluminum Company. Over the course of World War II, West Bend Aluminum Company would receive this award on six occasions. The first time the award was received it was in the form of a burgee or flag. Additional awards were in the form of stars to be added to the burgee. Each employee of the company also received an E Award certificate. Receiving this award was a great source of pride for the West Bend Aluminum Company, its employees, and the community.

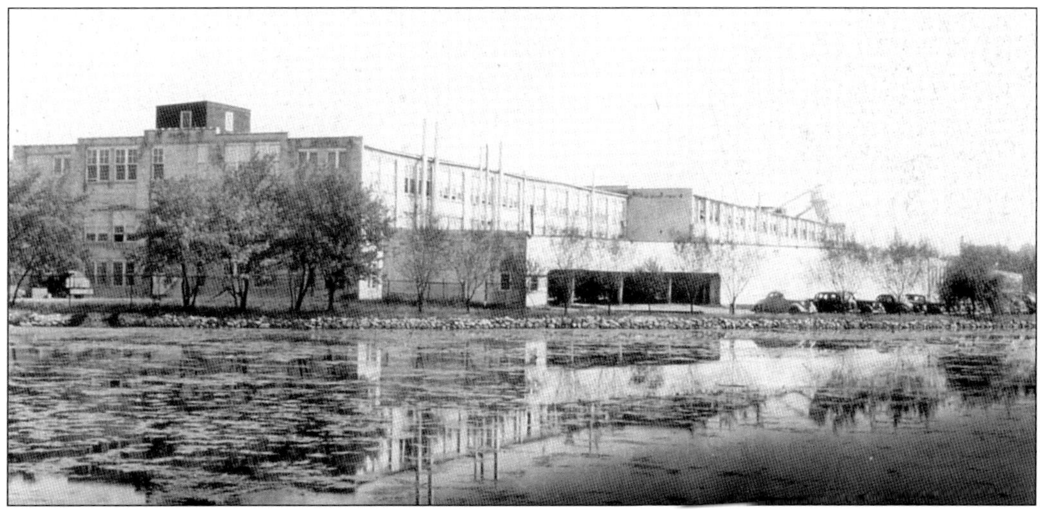
In 1944, West Bend Aluminum Company purchased the Kissel Industries plant in Hartford. This plant began as the Kissel Motor Car Company in 1906. The company manufactured automobiles under the name Kissel Kar. In 1931, the Kissel Motor Car Company was reorganized into Kissel Industries. This new company built products for Sears, Roebuck and Company and the U.S. government. By 1944, the plant had converted to manufacturing war items such as rocket casings under contract for the U.S. Department of War. West Bend Aluminum Company purchased the plant with an eye to the future of expanding its product lines. After World War II ended, West Bend Aluminum Company converted the plant to manufacture outboard motors. Motors under the names of Elgin and Shark as well as motors for Sears were produced.

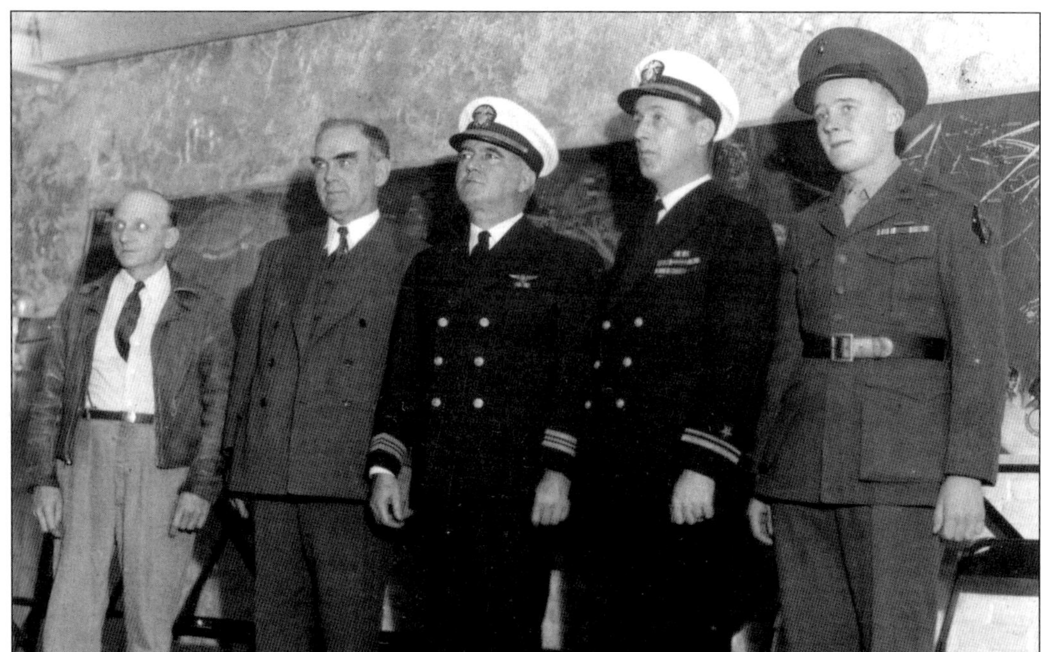

This photograph is from Rally Day at the Hartford West Bend Aluminum Company factory on September 30, 1944. From left to right the participants are George Ashenmacher, B. C. Ziegler, Lt. Comdr. J. F. McEndy, Lt. J. L. Palmer, and Pvt. Herman J. Christensen. George Ashenmacher was an employee at the Hartford plant and had three sons on active duty.

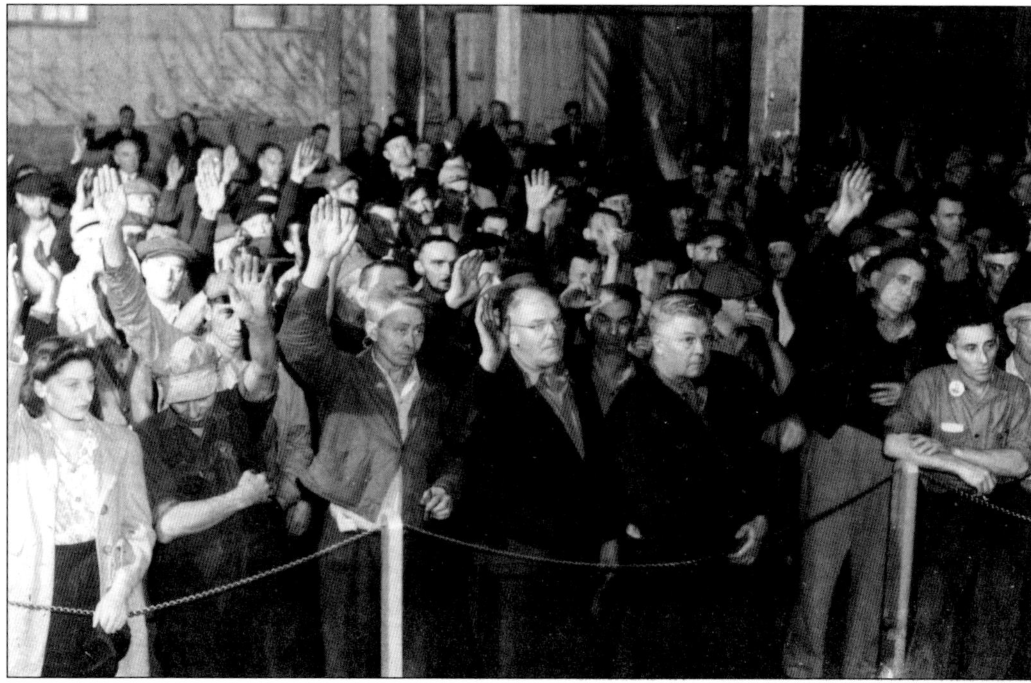

This photograph is a view of the crowd on Rally Day at the West Bend Aluminum Company's Hartford plant. Here employees raise their hands to indicate that they have close relatives in the service.

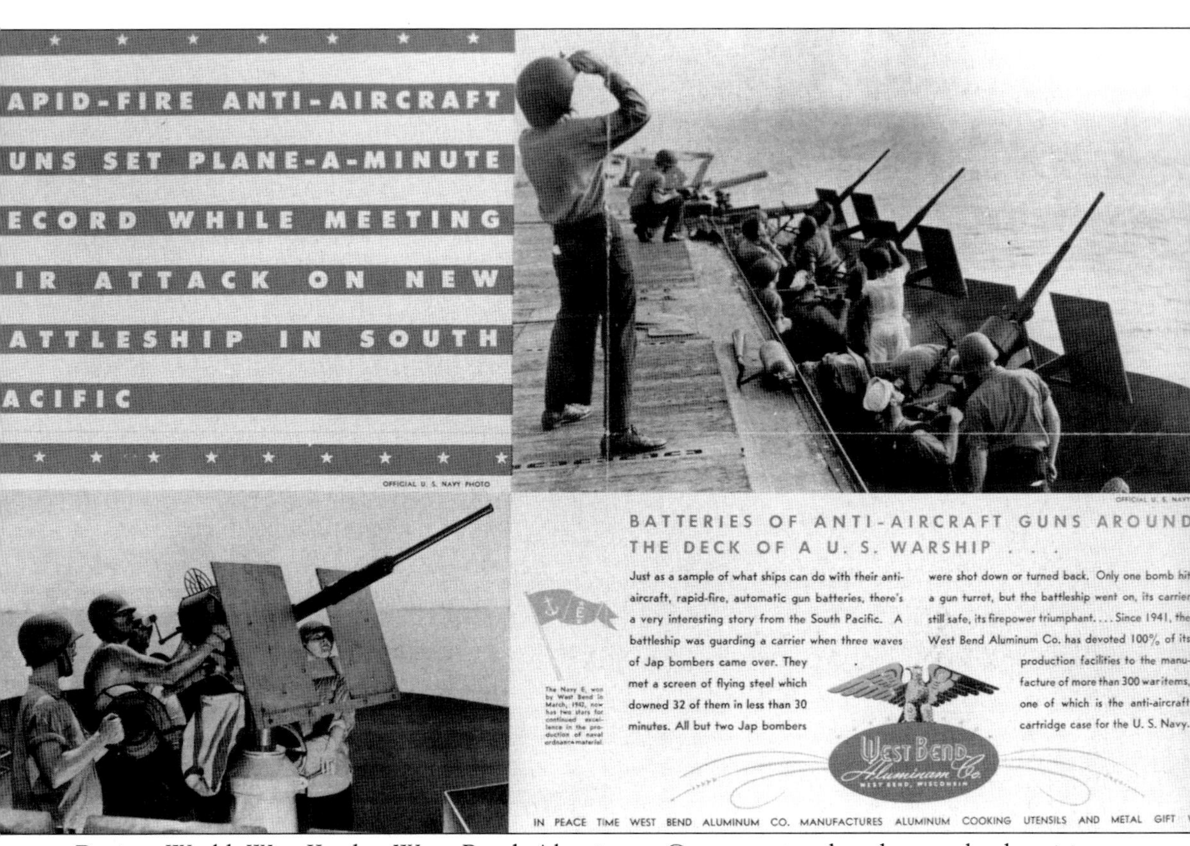

During World War II, the West Bend Aluminum Company produced several advertising campaigns to promote their patriotism and keep their name in the mind of consumers. This poster was one of those. Other advertisements included products and made it clear that production would continue on popular products such as the Trig teakettle after the war.

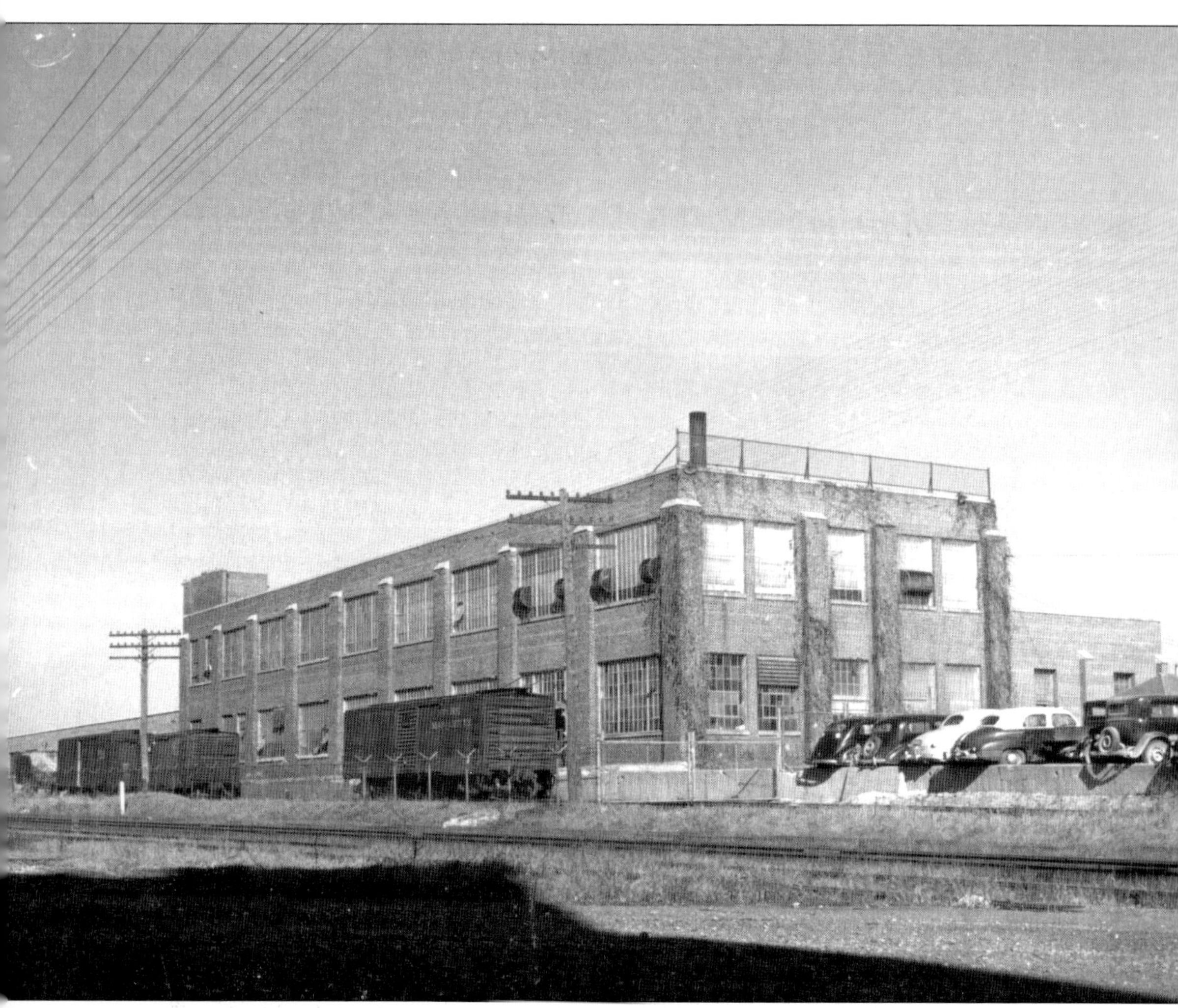

This photograph shows the Kewaskum Aluminum Company factory around 1944. Kewaskum Aluminum Company won a contract with the U.S. Department of War for experimentation and production of cartridge cases. The specifications demanded by the U.S. government were much higher than what the company had been accustomed. Kewaskum Aluminum Company worked on the development of steel cartridge cases to supplement the usual brass casing due to a brass shortage. On April 2, 1942, the company received the contract to manufacture 75-millimeter steel cartridge cases. The government expanded the program, awarding compliments from the U.S. Department of War and further contracts. The decision to sell the company came after the deaths of Adolph Sr. and Adolph Jr. Rosenhiemer and the discussion over the problems of conversion to peacetime production. The general consensus was to cease production while it was still profitable. J. O. Reigle and associates won the bid and the final transfer of ownership was made on August 6, 1945.

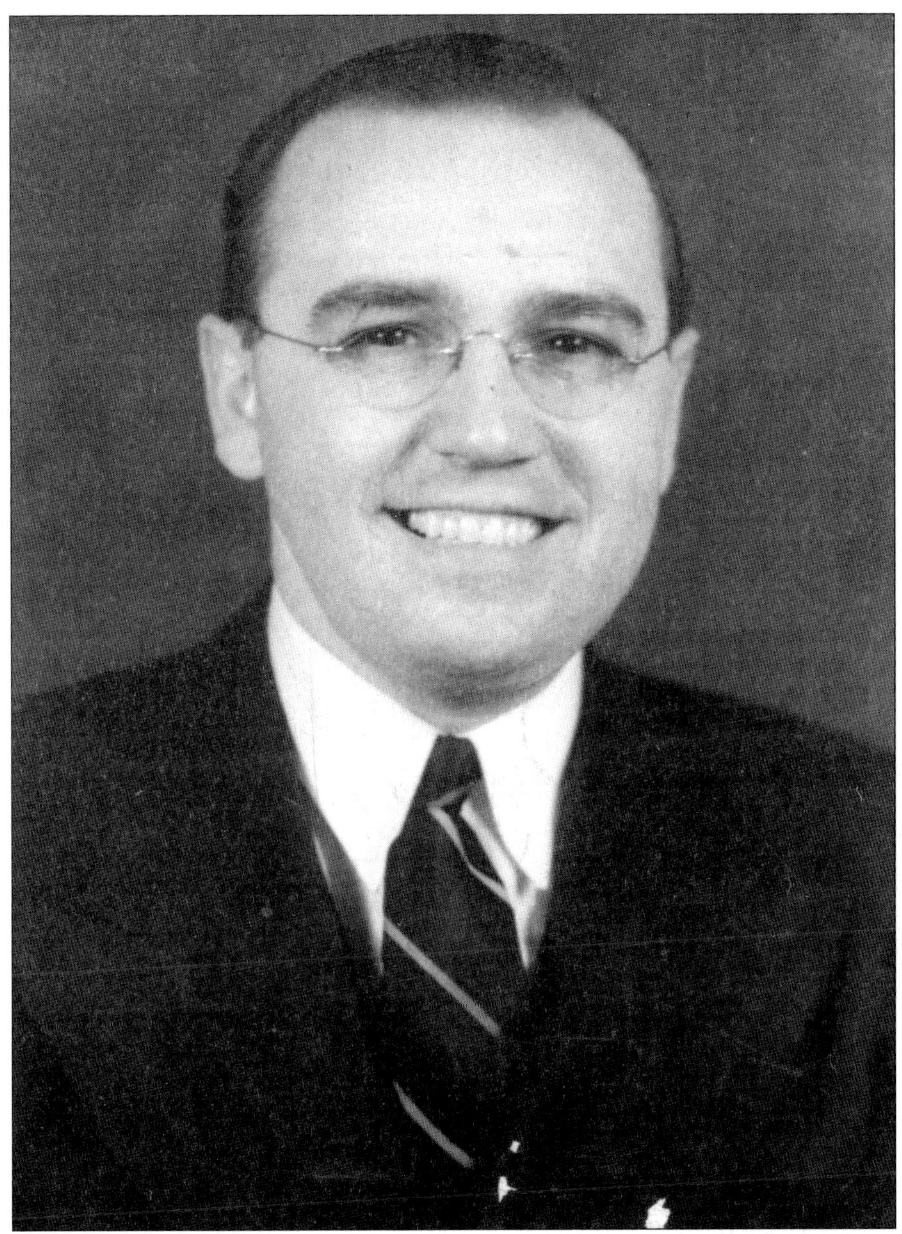

J. O. Reigle served as the head of Regal Ware Inc. from 1945 until 1965, when his son, James D. Reigle, was named president. J. O. remained active in the company and chairman of the board until his death in 1973. Born in 1898 in Leesville, Ohio, J. O. grew up on his family's 160-acre self-sufficient farm. While working to put himself though college, J. O. discovered his gift for salesmanship. After an enlightening business visit to a sanatorium in Battle Creek, J. O. became very interested in a new product, waterless cookware, and began his career in cookware sales. J. O. was so successful in his direct-sales approach that he was able to build and furnish a house for his family, paying for it in cash with his sales proceeds. J. O. prided himself on knowing his customer. Known as a dynamic salesman, innovator, and entrepreneur, as well as a humanitarian, J. O. was very interested in people, in particular his employees. He always remembered birthdays and anniversaries.

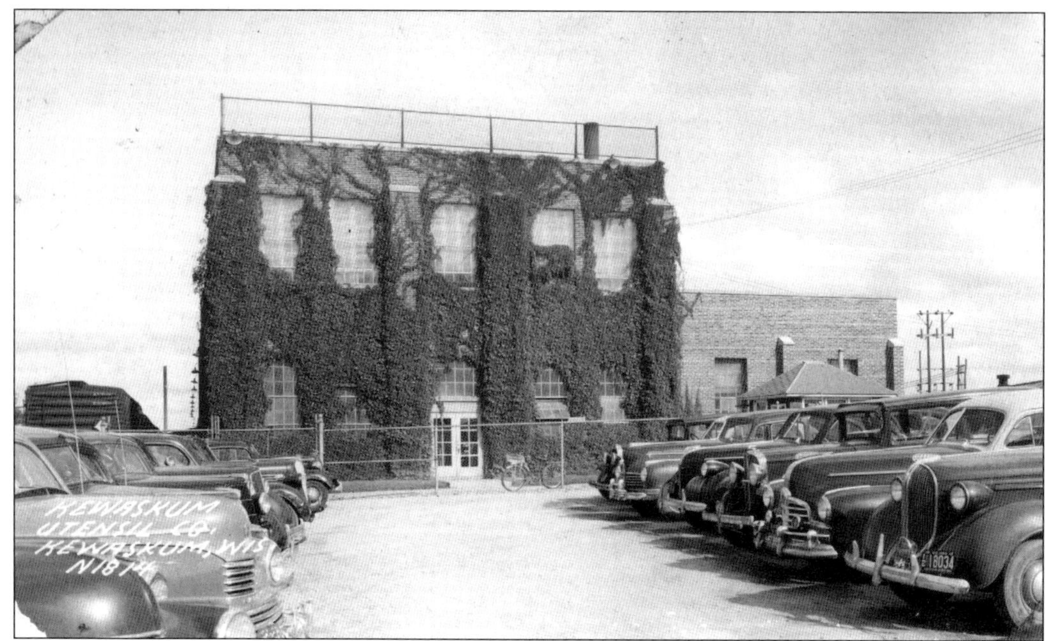

After the purchase by J. O. Reigle and associates, the Kewaskum Aluminum Company was renamed the Kewaskum Utensil Company. This photograph was taken shortly after the transfer of the business in 1945. In 1951, the company changed its name to Regal Ware Inc., taking on the product trade name of Regal Ware.

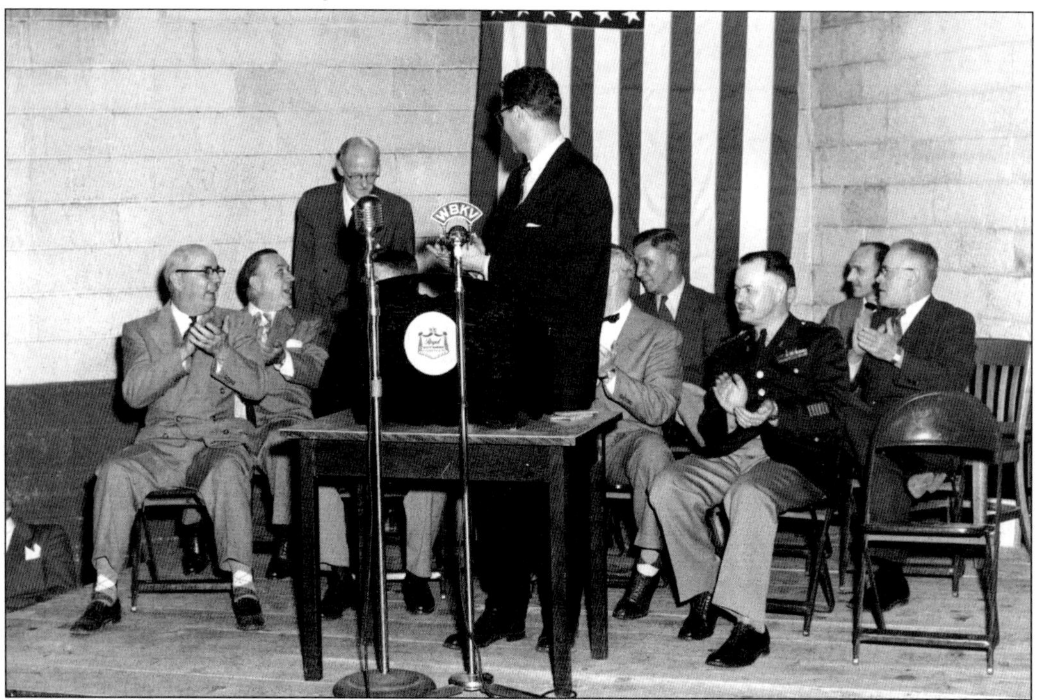

With the onset of the Korean War in 1950, the government again began awarding numerous defense contracts. Here military and Regal Ware Inc. officials attend a ceremony awarding a government defense contract to the company in February 1951.

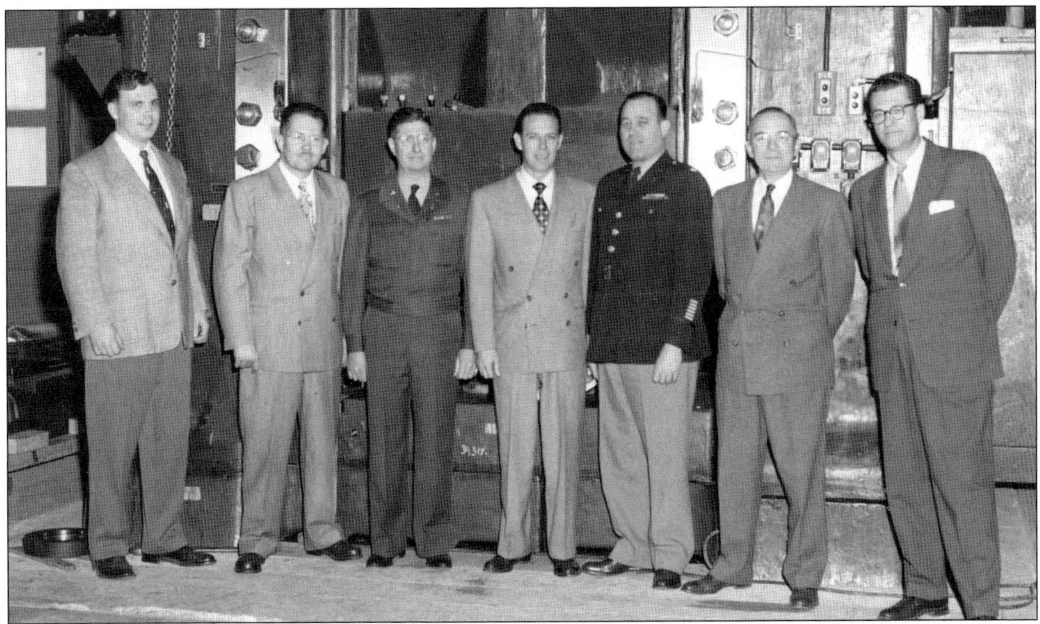

This photograph of U.S. Army inspection officers was taken in February 1952. The group is standing in front of the Bliss No. 33 knuckle-joint press, which was installed specifically for the army contract. An entire new wing of the factory was constructed to house the project. Although the original outlay and design was by Regal Ware Inc. the company was repaid by the army upon completed installation of this main press.

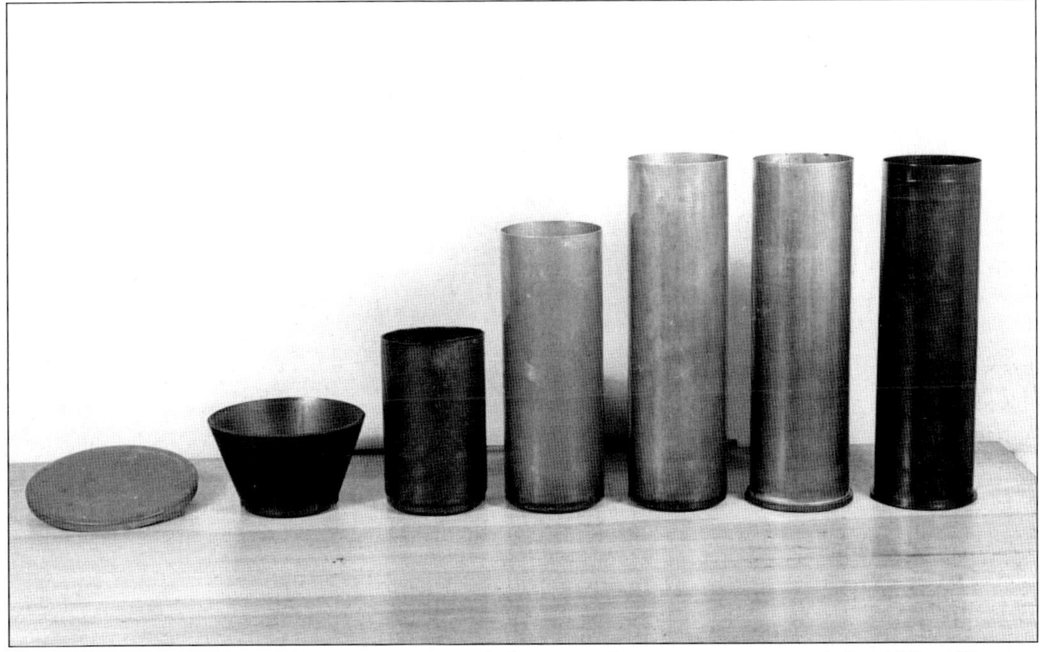

This photograph shows the seven stages of the manufacturing process of the 105-millimeter steel shell casing manufactured by Regal Ware Inc. The blank seen on the far left is seven and three-eighths inches in diameter and over one half inch thick. Before this process, designed by Regal Ware Inc., several additional draws would have been necessary.

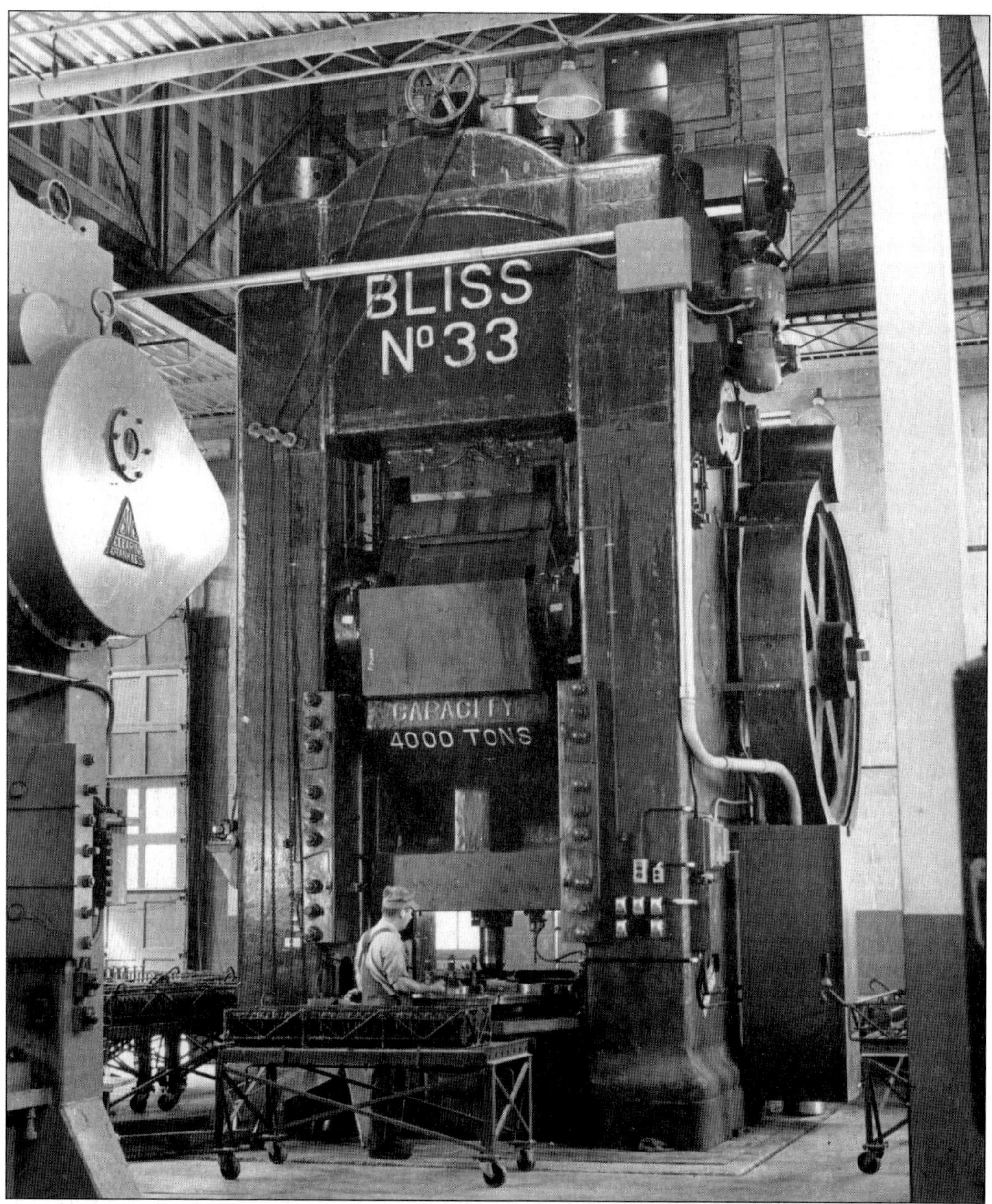

This photograph shows the Bliss No. 33 knuckle-joint press in use at Regal Ware Inc. in 1952. At the time, this was the largest knuckle-joint press in the world. At 34 feet tall and weighting 610,000 pounds, the press was able to exert eight million pounds of pressure. It took nearly a year to construct the press. It was engineered specifically for the production of 105-millimeter steel cartridge casings. Before the introduction of the Bliss press, a separate step was needed for the initial shaping of the cup and heading along with five separate draws. The press allowed the first cupping and heading to be combined and reduced the number of draws to three. This greatly sped up the production time and in turn lowered costs.

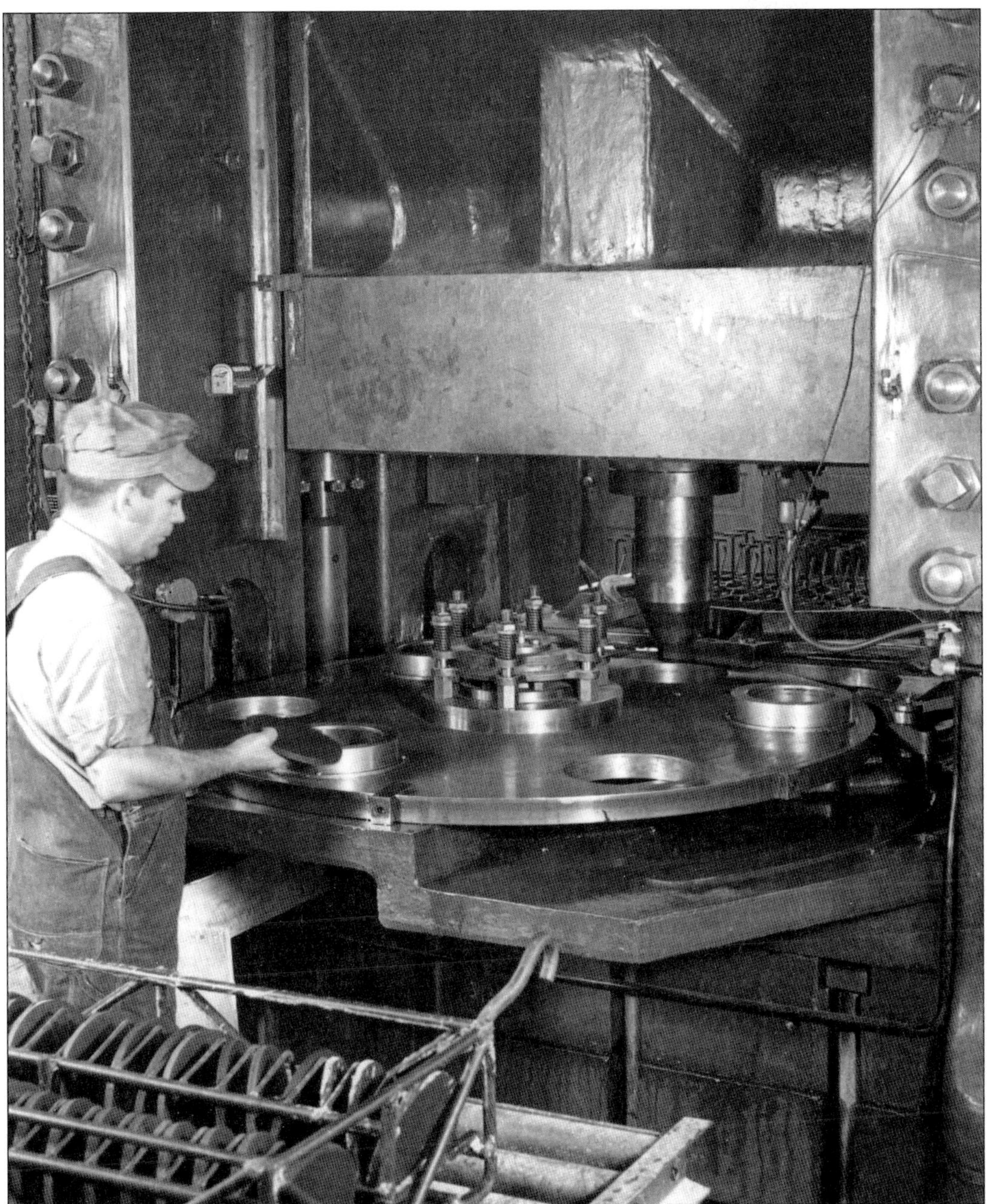

Regal Ware Inc.'s Bliss press had a six-station table for the first step of the 105-millimeter shell production. Before being altered, the steel blanks were immersed in a phosphate bath, rinsed in water, and then submerged in a lubricant. This created a coating that allowed deeper draws and fewer steps to the manufacturing process. In this photograph, the first step creates the cup shape needed for the shell. Ten of these cups could be made in one minute. The tapered cup measured three and a half inches deep and six and five-eighths inches in diameter across the top. The cups would drop down a chute after being shaped, on to the next step of the process.

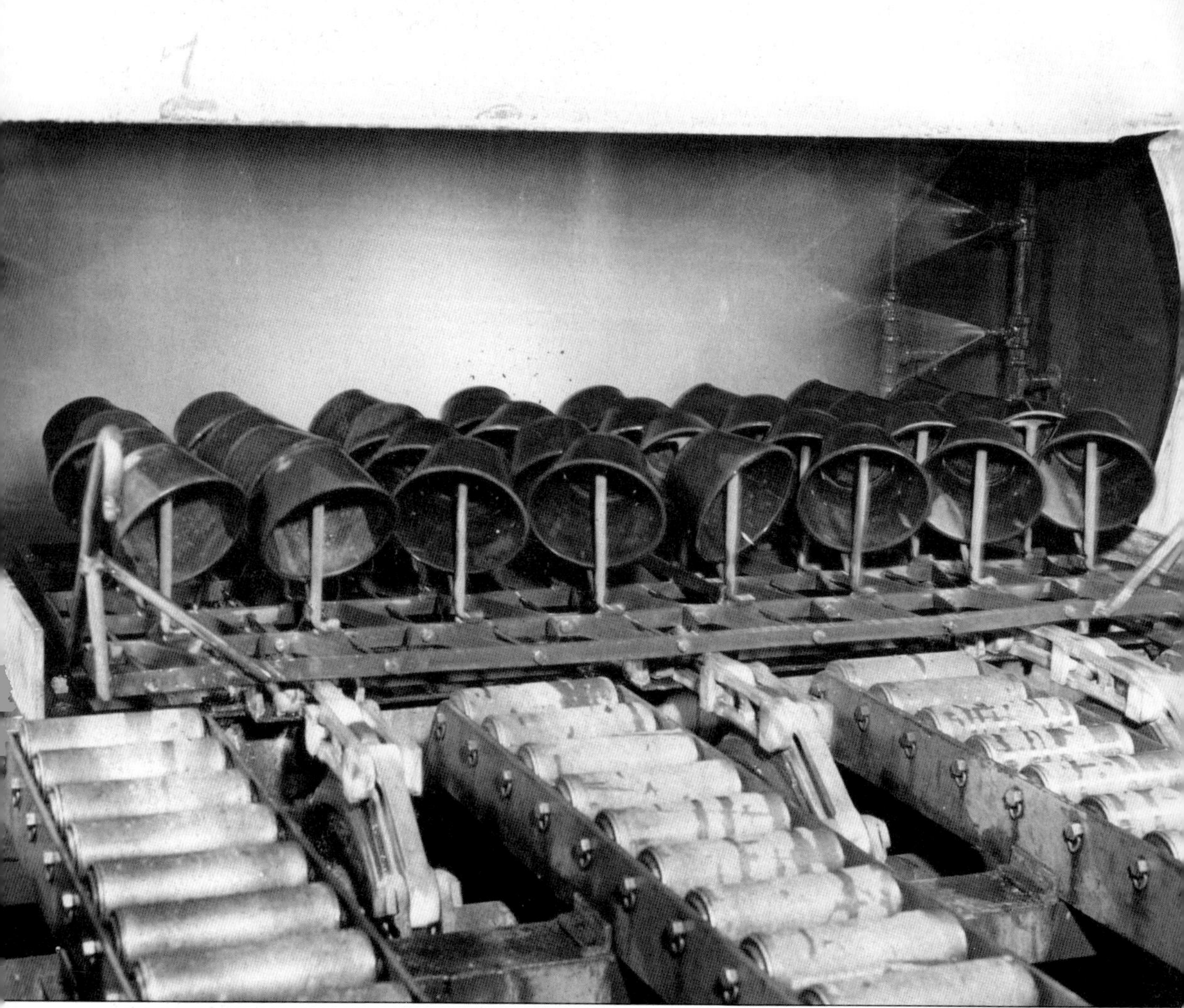

Once the cup was formed, the 105-millimeter shell moved into the annealing oven for 30 minutes. After exiting the oven, the shells were sprayed with water to cool them down and remove any loose particles. In this photograph, the shell cups sit in their cooling shower before moving on to a sulfuric acid bath and a second coat of lubricant. The casings then moved on to the first draw. This draw produced a cylinder eight inches deep and four and a half inches in diameter. The Regal Ware Inc. presses could process 600 draws per hour. After the first draw, the annealing, washing, and lubricating process was repeated.

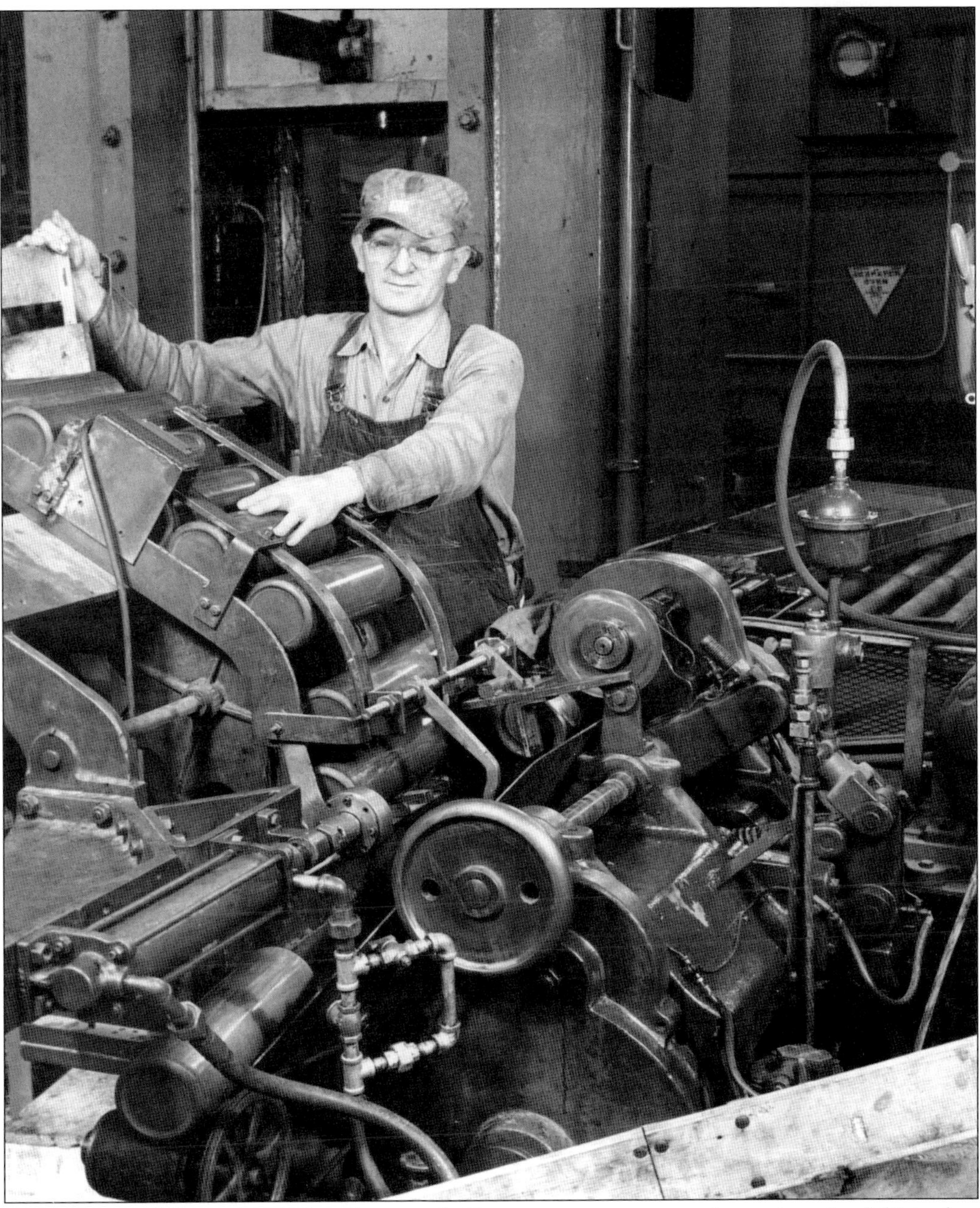

After a second draw, the 105-millimeter shell casings were trimmed to a length of 12 inches and 13½ inches deep. This photograph shows employee Chris Kober feeding the shells though a circular cutter to achieve this length. The cut was made at a 45-degree angle so there were no chips or burrs. The casing then continued though a third and final draw, trim, and heading. The final length of the casing was 14⅞ inches. The head of the casing was created and an inspection of the head and body was carried out before the final trimming and a primer hole was drilled.

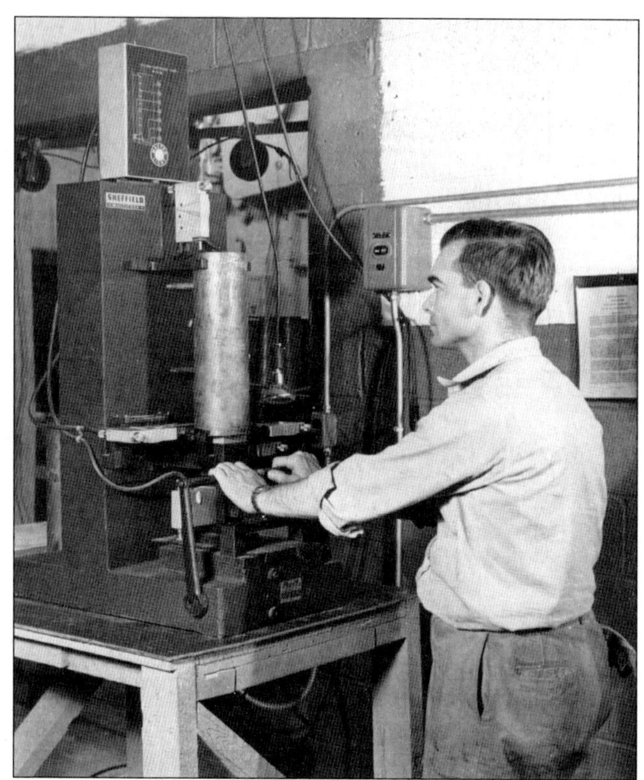

In this photograph, the finished 105-millimeter shell casing is having its final inspection. Here Clarence Reigle uses the Sheffield Milticheck gage to inspect seven dimensions of the casing at once. Signal lights let the operator know if the casing passed tolerances. Next the shell was stamped with the marking press, cleaned, and painted. Both the inside and outside of the casing were painted with an orange phenolic varnish that looks brown when dry. The casing was dried in an oven and underwent one final visual inspection before being packed and shipped.

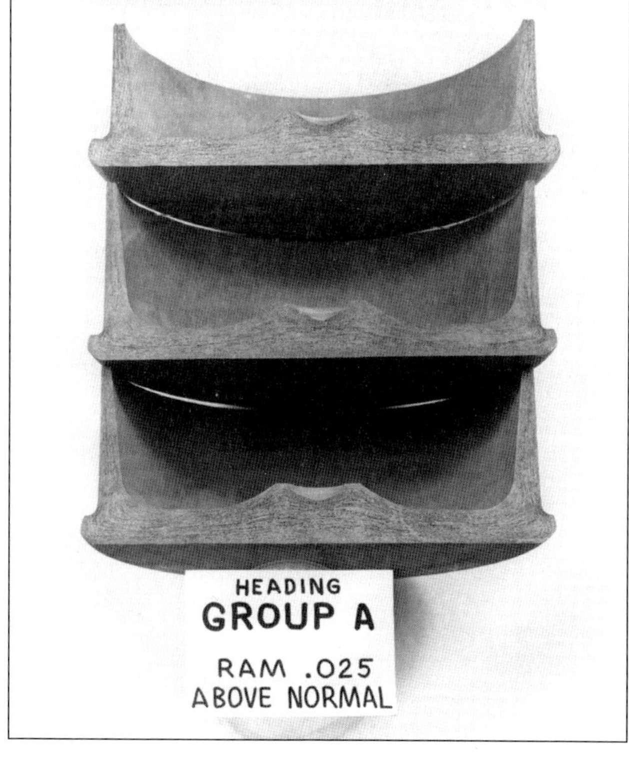

Seen here is a cross section of three steel 105-millimeter shell casing produced by Regal Ware Inc. This shell was one of the tests done to perfect the heading process. Notice how thick the base of the casing is when compared to the sides.

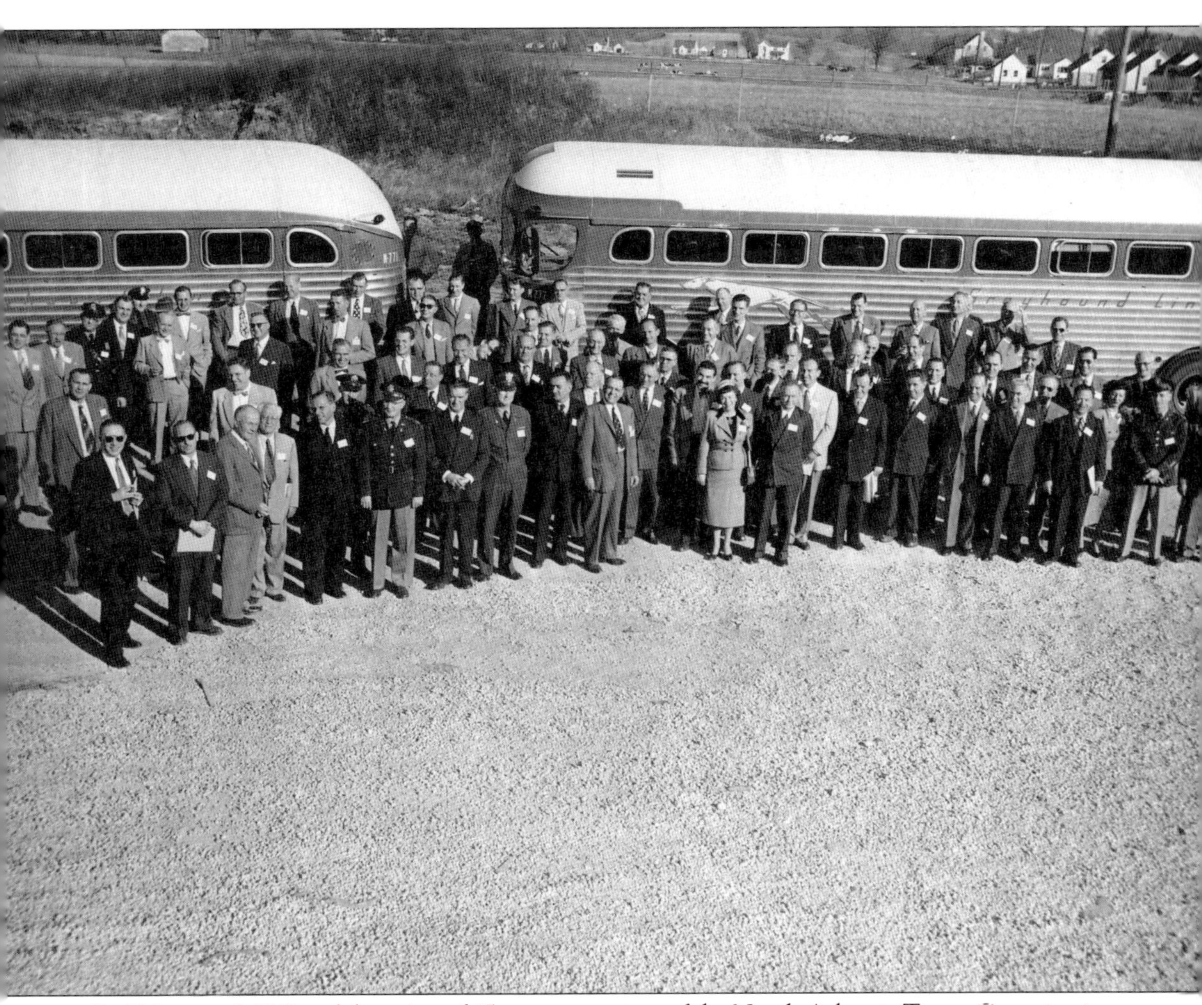

On February 6, 1953, a delegation of 45 representatives of the North Atlantic Treaty Organization visited Regal Ware Inc. This photograph shows the group in the parking lot of Regal Ware Inc. Eleven countries were represented and Regal was one of only four private industries in the United States visited by the committee. The tour was part of an effort to increase ammunition production in Allied countries. The group's goal was to study production techniques and exchange experiences to better their own production facilities. At Regal Ware Inc. the visitors learned of the development and production of steel cartridge casings.

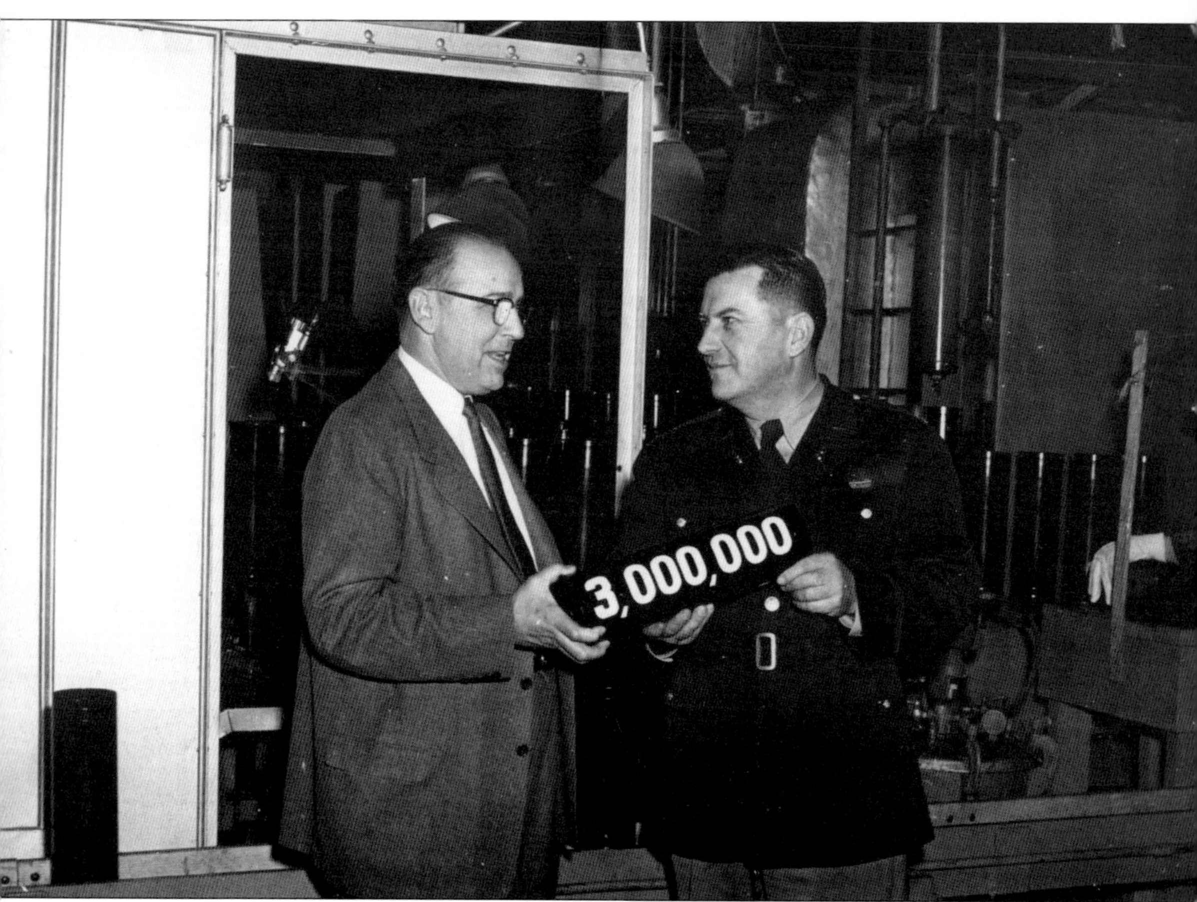

This photograph shows the ceremony commemorating the three millionth 105-millimeter shell produced by Regal Ware Inc. under defense contract to the ordnance department of the U.S. Army. The presentation took place on December 20, 1954. Here company president J. O. Reigle presents the shell to Col. L. F. Stangel, chief of the Chicago ordnance district. The presentation took place at the Kewaskum factory, and a tour of the facility followed.

Three

The Boom Years 1950–1980

The end of World War II saw the return of peace and stability to the country and returned manufacturers to consumer production. A new market appeared for housewares companies, fueled by the evolution of the modern family. Marriage rates skyrocketed and companies cashed in with appliances and gift sets for the booming bridal market.

Newlyweds bought into the flourishing suburban housing market as a comfortable and safe place to raise a family. Within these homes, the kitchen was designed to be the center of family life, and new appliances and cookware were needed to furnish them. New materials and technology popularized stainless steel, porcelain over aluminum, and plastics in turn allowing design innovations. Nonstick coatings rejuvenated cookware sales while electricity in the household encouraged innovations in small electric appliances. Items went from being purely functional to being aesthetically pleasing. Color became a factor in designs due to the properties of these new materials. Plastics and anodized aluminum in reds, yellows, greens, pinks, and blues ornamented the home in the 1950s.

The mid-20th century found America facing great political and economic change. These changes spurred housewares manufacturers to create new products and designs to fit new lifestyles. Women sought further education and filled more jobs, making up 50 percent of the workforce by 1980. Inflation required the support of two incomes and women relied more on "Mr. Mom" to help with the kids and household tasks. Housewares products reflected the convenience demanded by busy families. Nonstick surfaces, small electrical appliances, and comfort items such as humidifiers all became popular.

In the 1960s and 1970s, product research studies showed the color trends leading to earth tones. Popular colors included flame, harvest gold, and avocado. Home entertaining also became popular in these decades, and the small–electric appliance industry expanded to fit the demand.

Advancements in plastics made appliances lighter, damage resistant, and were combined with glass to produce microwaveable cookware. Purchasing markets changed, and department stores stocked hundreds of prepackaged appliances. The colorfully packaged appliances helped inform the consumer and acted as the main selling agent.

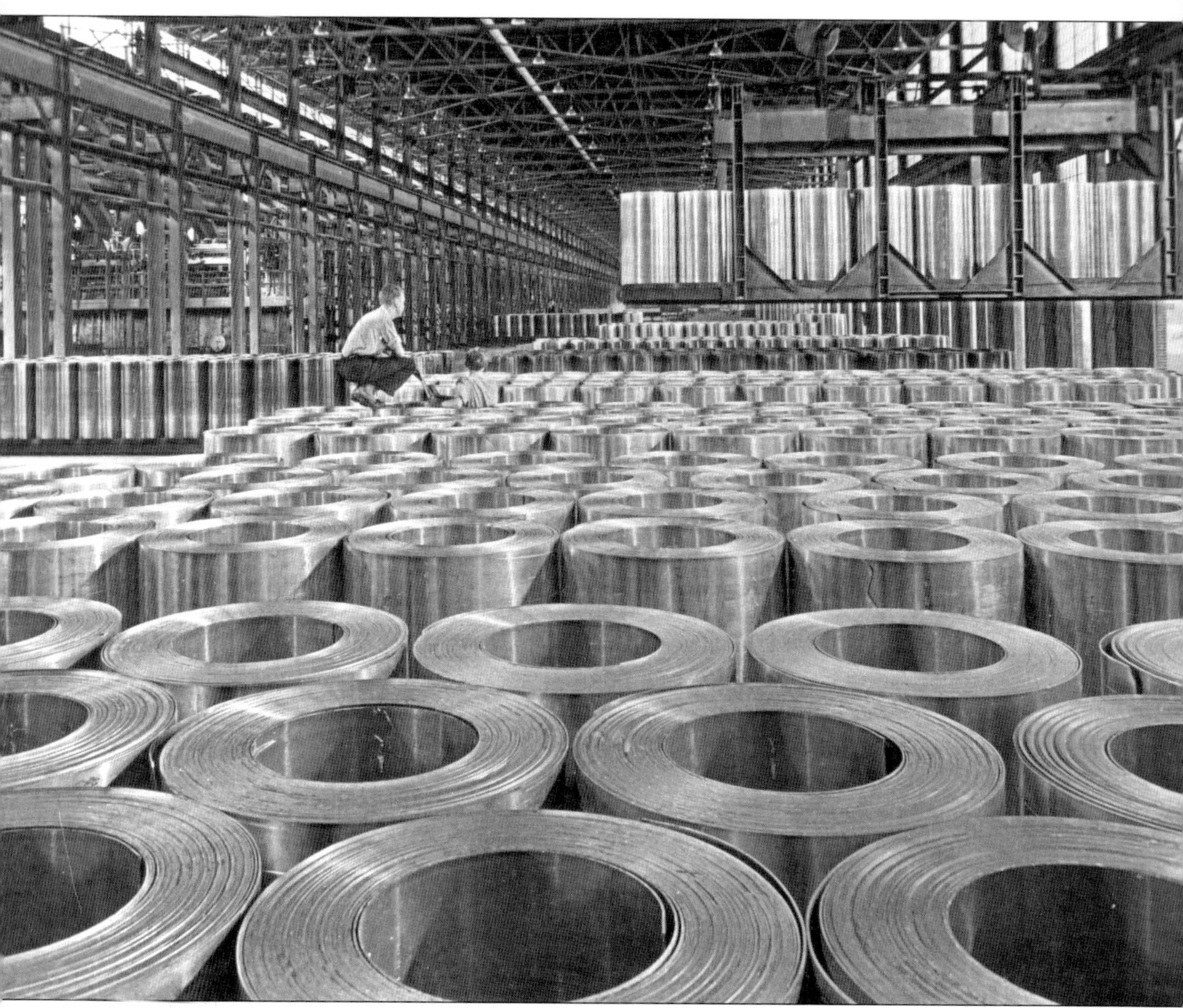

Rolls of sheet aluminum wait for processing in a warehouse of the West Bend Aluminum Company in 1951. West Bend Aluminum Company produced its first set of stainless steel cookware in January 1951. Stainless steel was easier to clean and more sanitary than other cookware metals.

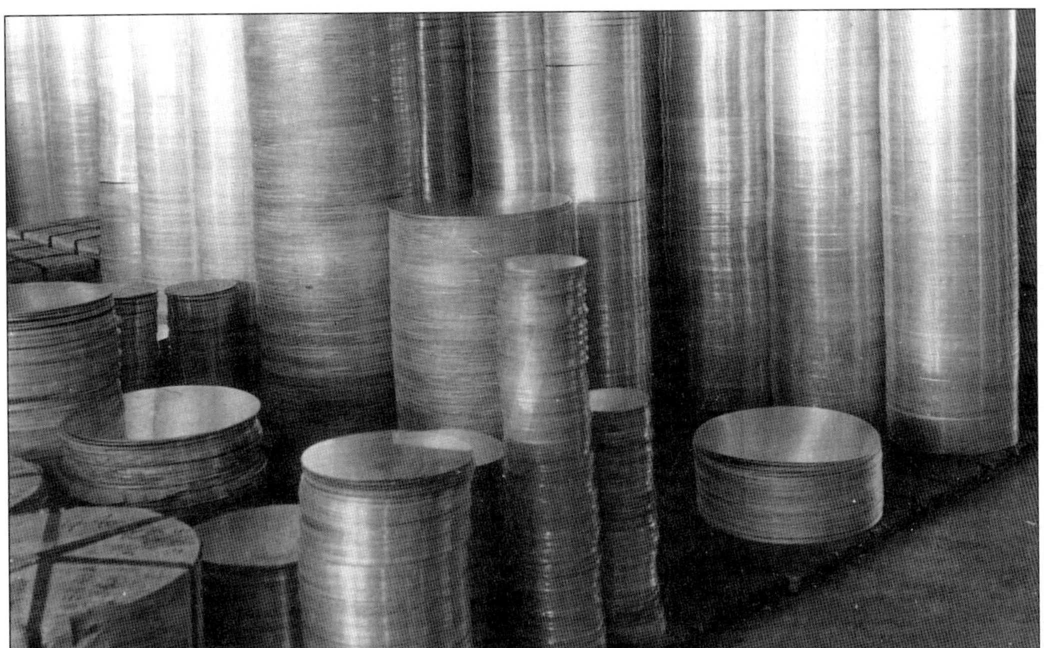

Stacks of aluminum blanks wait in the Metal Room at the West Bend Aluminum Company around 1951. The majority of aluminum was purchased in this form. Because most utensils were round and conformed to standard sizes, pre-shaped aluminum was the most economical way to proceed. It reduced the need for rolls of aluminum, machinery, and waste aluminum.

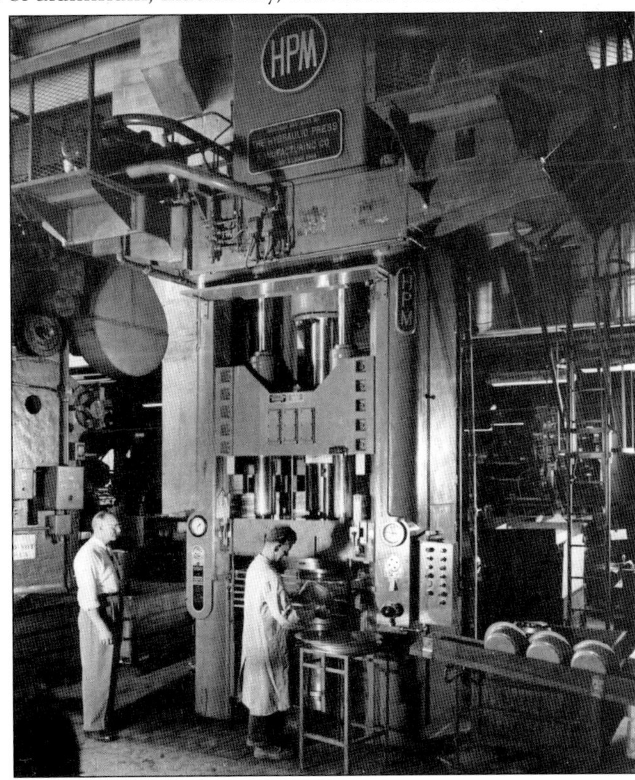

This photograph shows the 300-ton double-acting hydraulic draw press at the West Bend Aluminum Company around 1955.

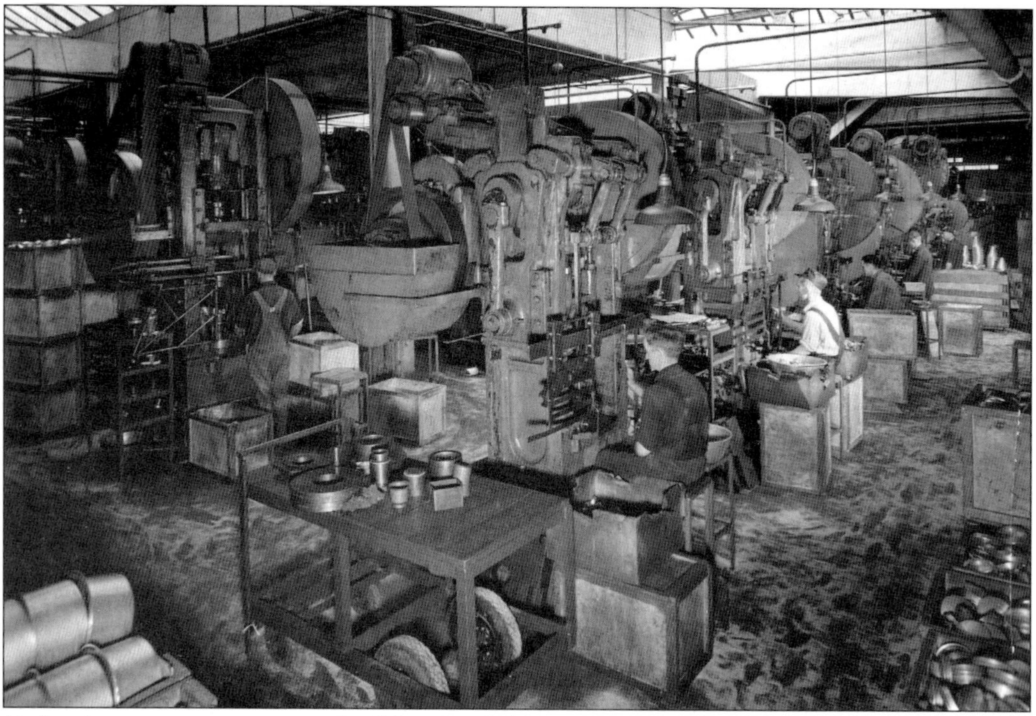

Hydraulic presses are seen in the West Bend Aluminum Company Press Department around 1950. Hydraulic presses were used for forming heavy-gauge metals.

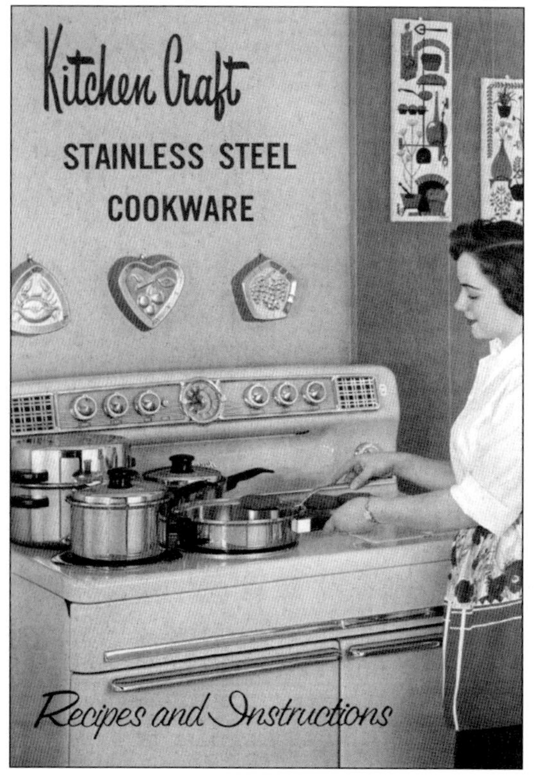

This image is from the instruction and recipe book that came with a set of Kitchen Craft cookware in 1957. Kitchen Craft, which was made by the West Bend Aluminum Company, was a heavy-gauge stainless steel line. Kitchen Craft was one of the waterless cookware systems from the company.

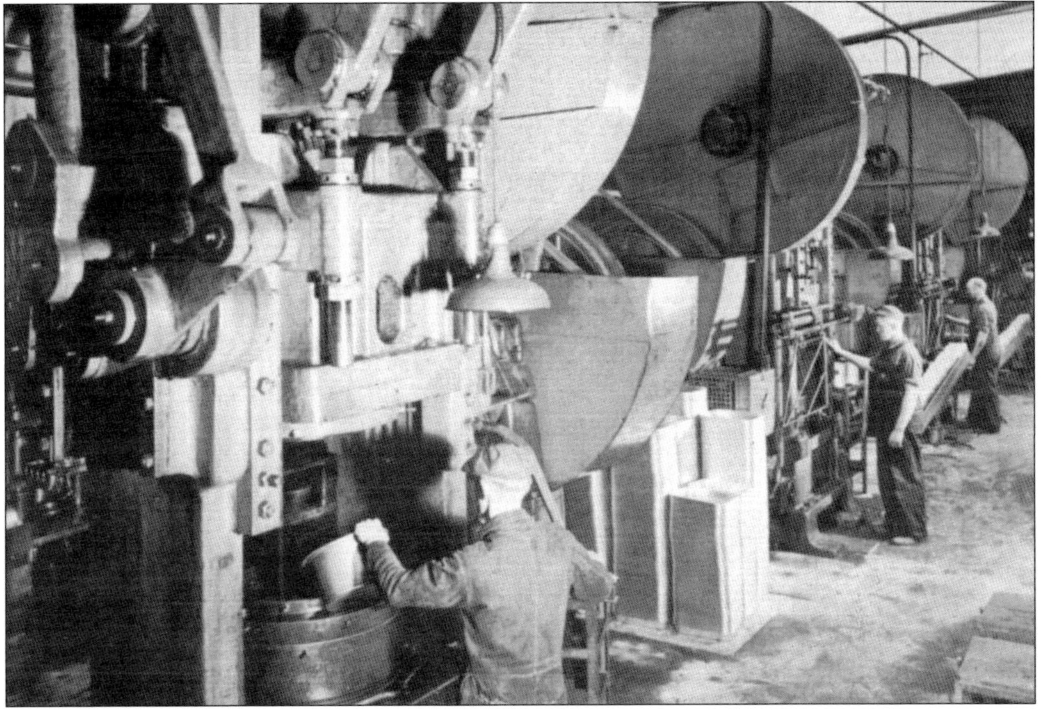

This photographs shows part of the West Bend Aluminum Company's Draw Press Department. These smaller presses were used to make smaller and lighter-duty pieces.

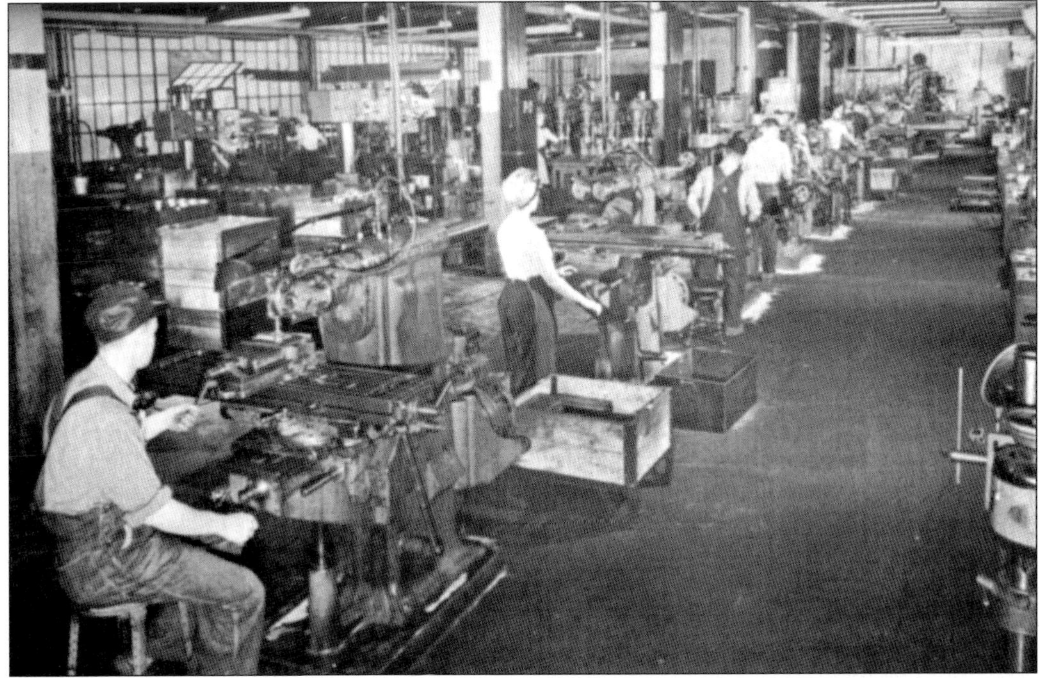

A row of milling machines is seen in operation at the West Bend Aluminum Company around 1950. Skilled men and women used these machines to create parts. These parts were in turn used in the machines that would manufacture the product.

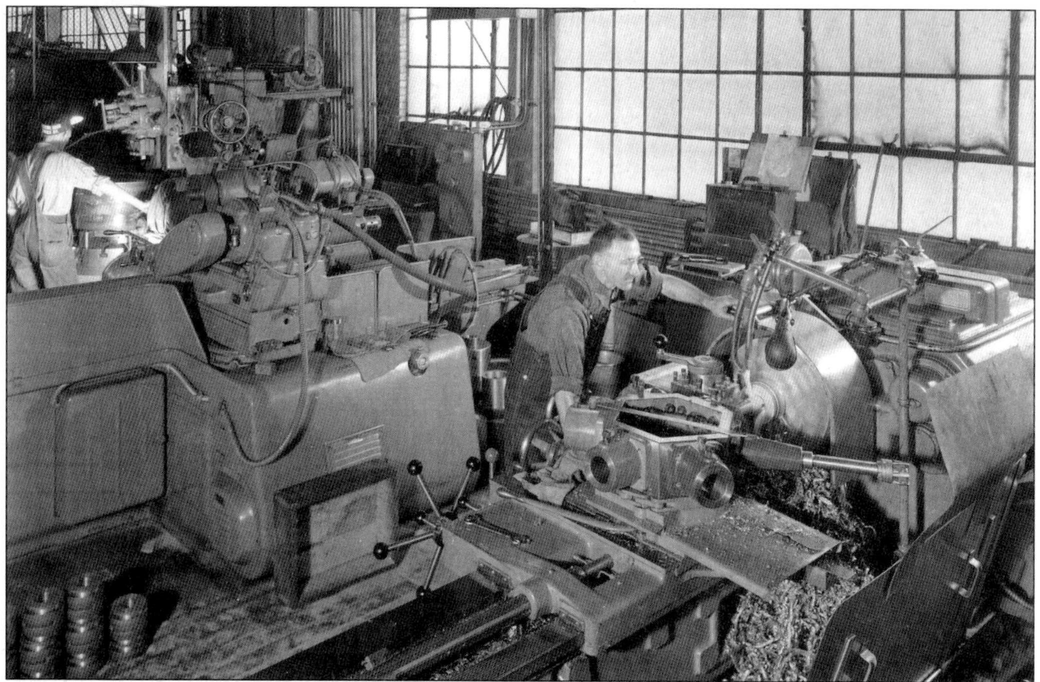

This 1950s photograph shows the lathe section of the machine shop in action. Engine-driven vertical turret lathes and horizontal turret lathes were used to make parts for the machines that would actually create the products.

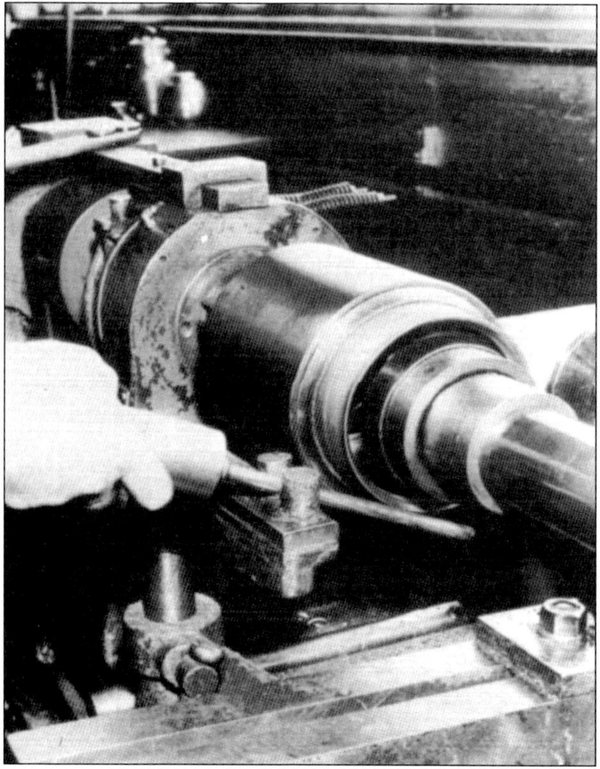

This photograph shows a spinning lathe in action around 1950. Operators at the West Bend Aluminum Company use the machine to form product shells into their final shape.

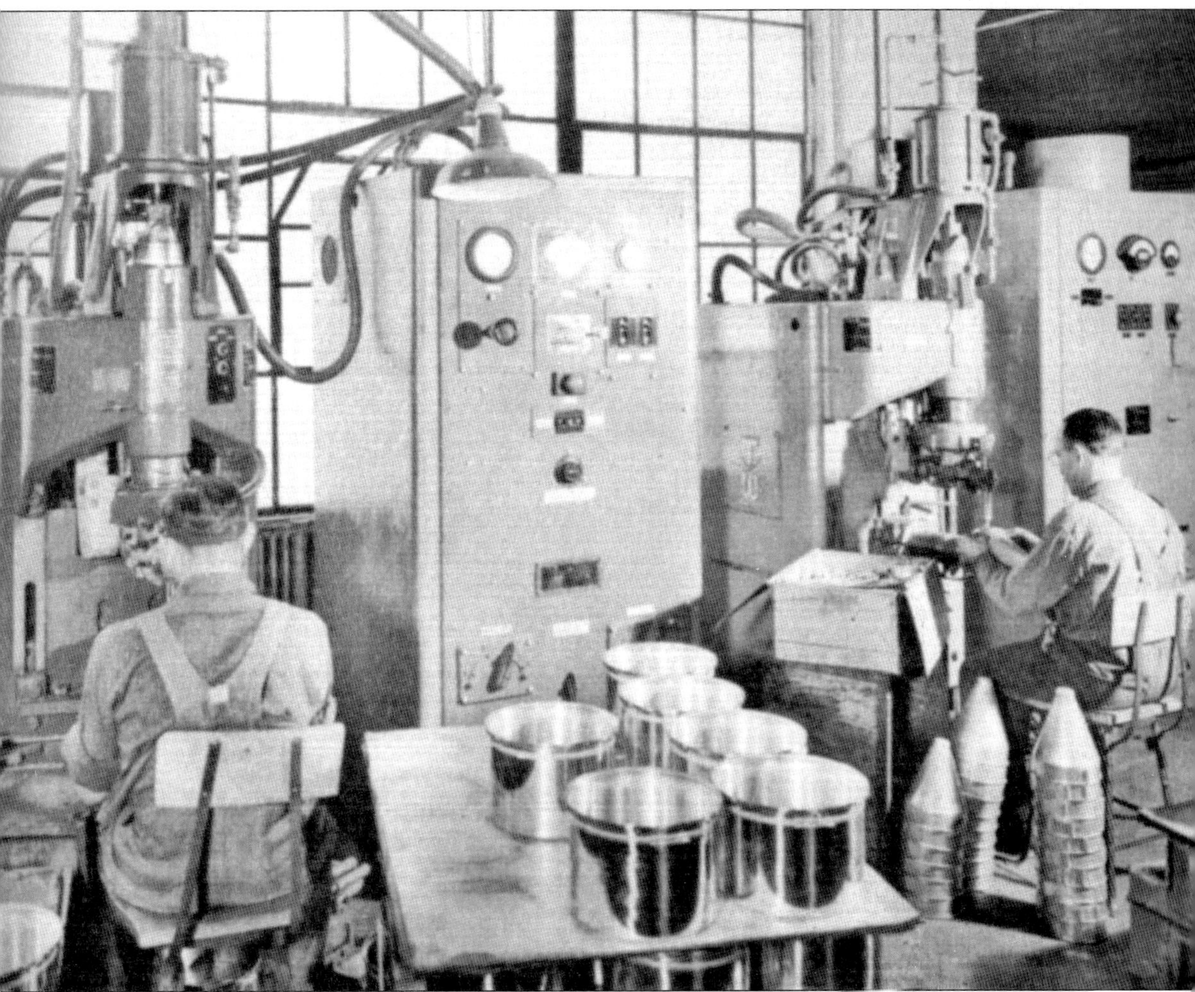

This photograph shows the spot welding equipment of the West Bend Aluminum Company in use around 1950. State of the art for the time, the Hi-Wave welding equipment could weld a wide range of thickness of aluminum as well as steel. The welders could complete up to 1,100 aluminum cooking utensil jobs per hour, depending on the job.

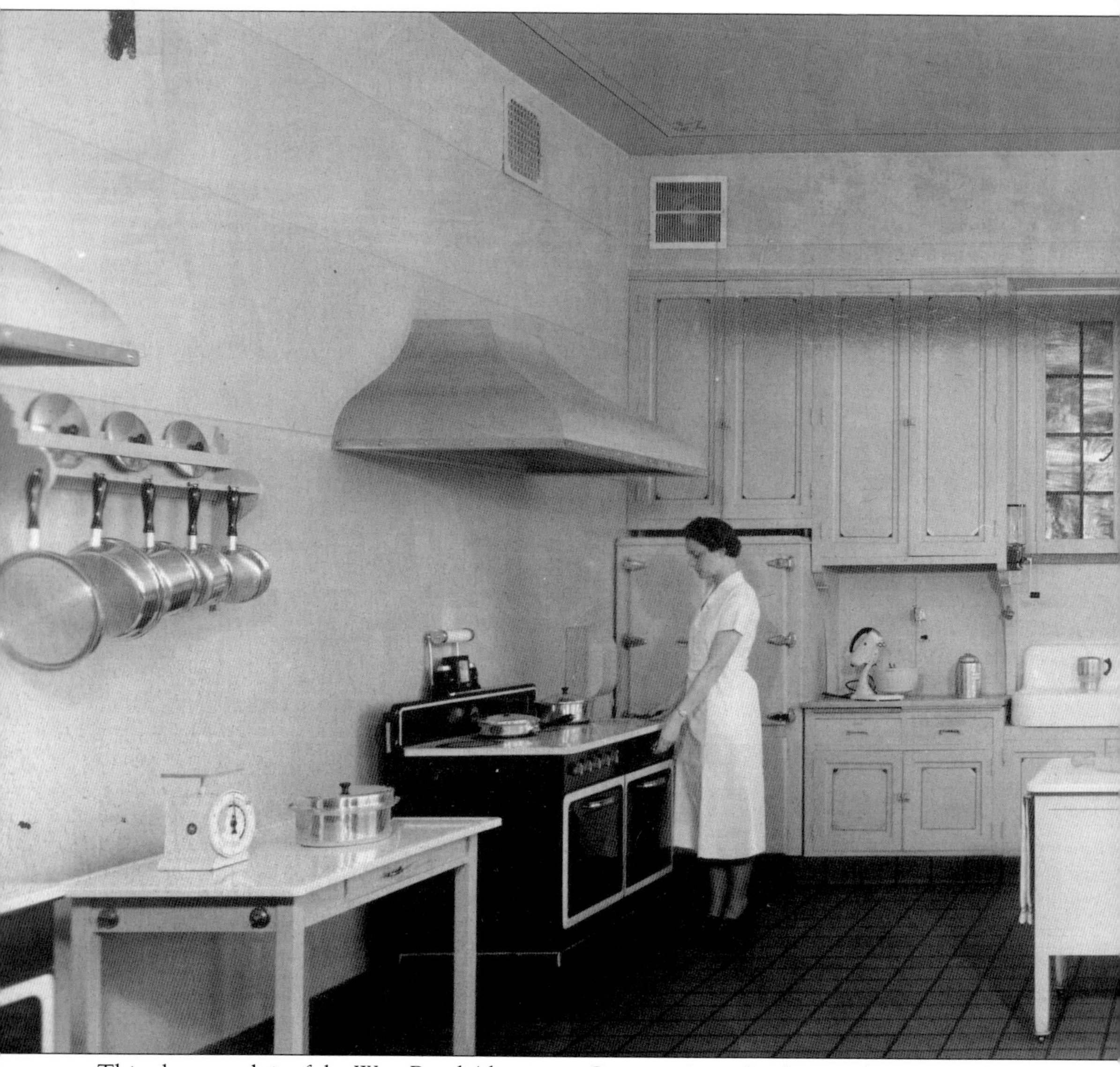

This photograph is of the West Bend Aluminum Company's test kitchen in the 1950s. Although the same room was used, it was constantly updated. The icebox from the earlier kitchen has been replaced with a refrigerator and freezer. The stove has also been replaced with a modern model. Note the early electrical appliances on the back counter, a sign of the times. The kitchen

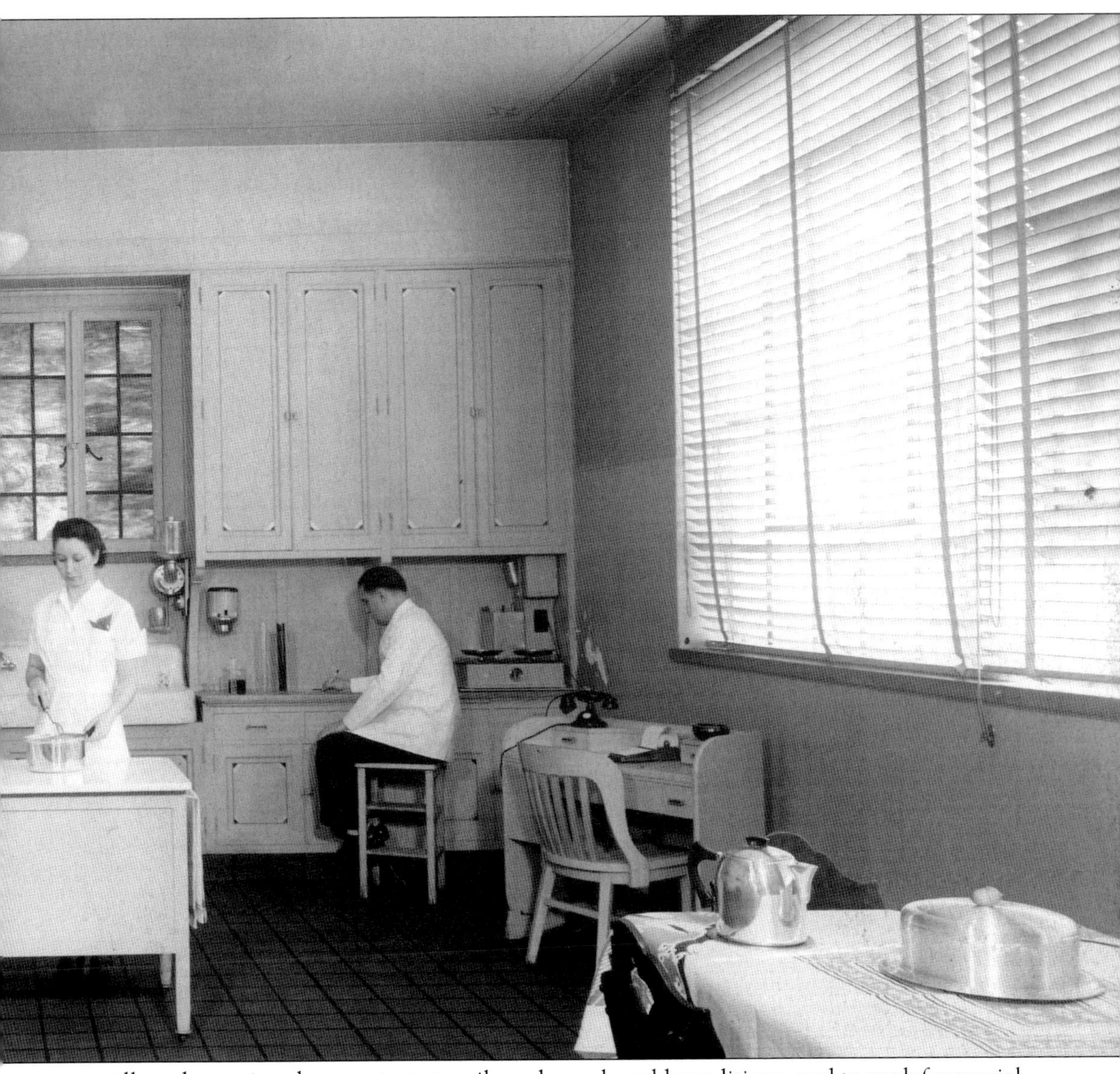

was still used to train salesmen, test utensils under real-world conditions, and to cook for special guests of the company. The test kitchen workers also created recipe books and instructions for use with specific utensils.

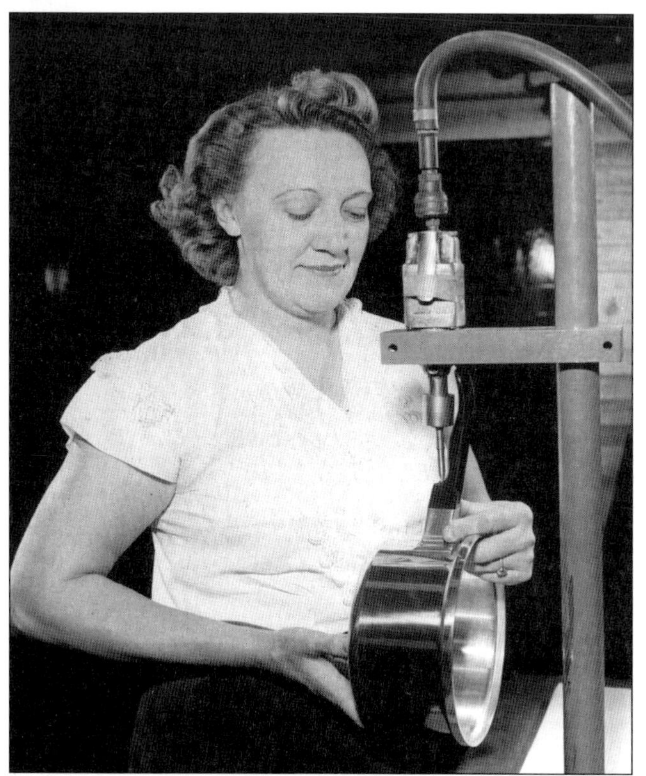

Here a worker uses an automatic screwdriver to attach the handle to a Kitchen Craft pan around 1955. The Kitchen Craft line was a cookware system that promoted itself as a healthy way to cook. Just like the Waterless Cooker, the Kitchen Craft system was able to produce meals with little or no water added because of a tight-sealing lid. Originally made from aluminum, the product began being made of stainless steel in the mid-1950s.

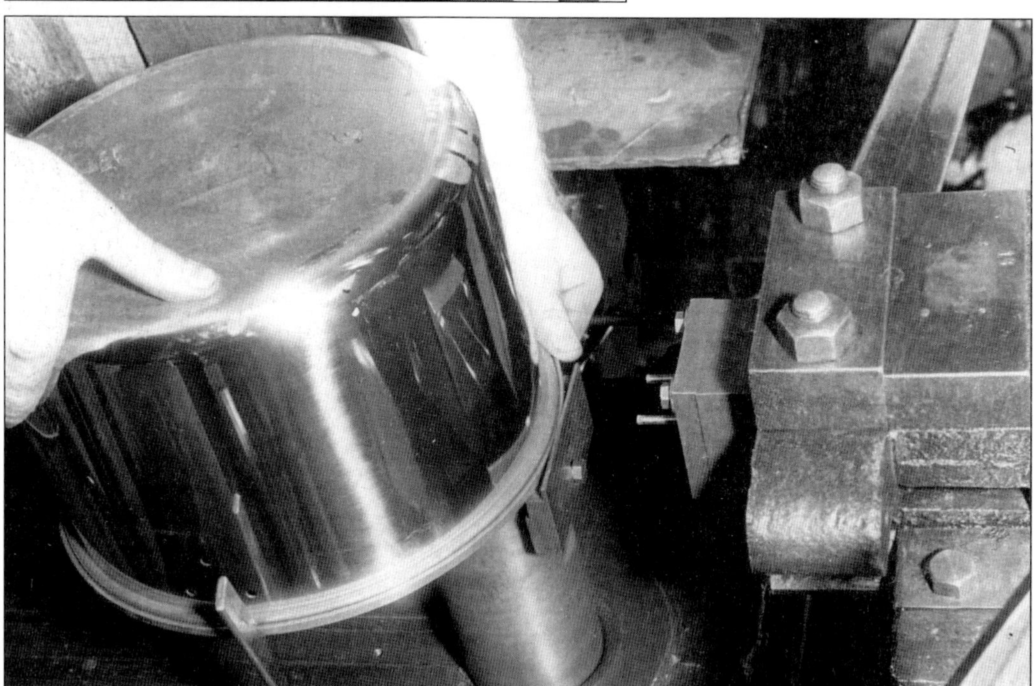

A punch press makes holes in a Kitchen Craft pan around 1955. The punch press was used to stamp holes used to attach handles, hinges, knobs, and other accessories. The punch press was also used to stamp holes in the filters for the drip coffeemakers.

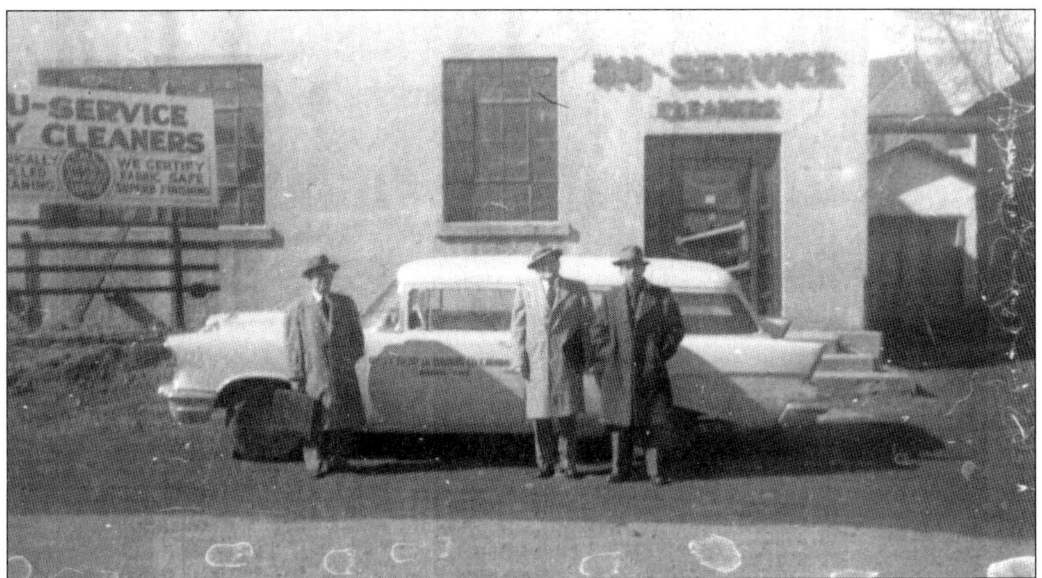

This photograph shows the first factory of West Bend of Canada around 1953. The company manufactured outboard motors above the Nu-Service Cleaners from 1952 to 1958. The men standing in front of the company car are unidentified but thought to be the Canadian sales representatives.

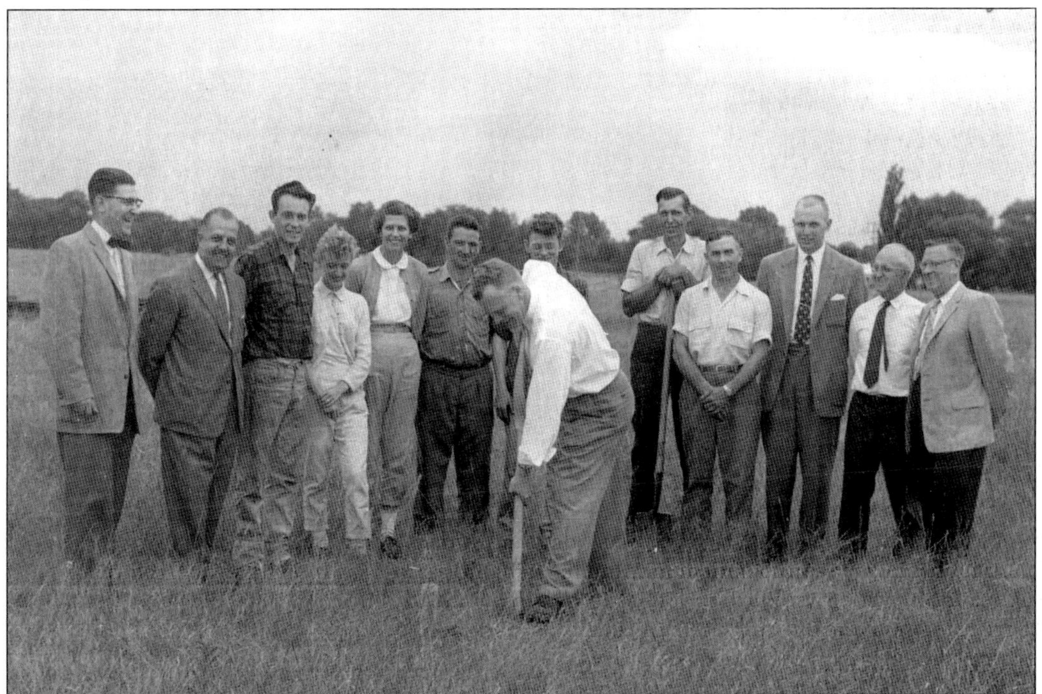

A group gathers for the groundbreaking of the West Bend of Canada factory on John Street in Barrie, Ontario. In August 1957, staff and employees gathered to watch the president of West Bend of Canada, Jerry Emberson, break ground for a new factory. The company had expanded into Canada in 1952 in makeshift quarters. The building of the factory proved the company was strong enough to make it as an international entity.

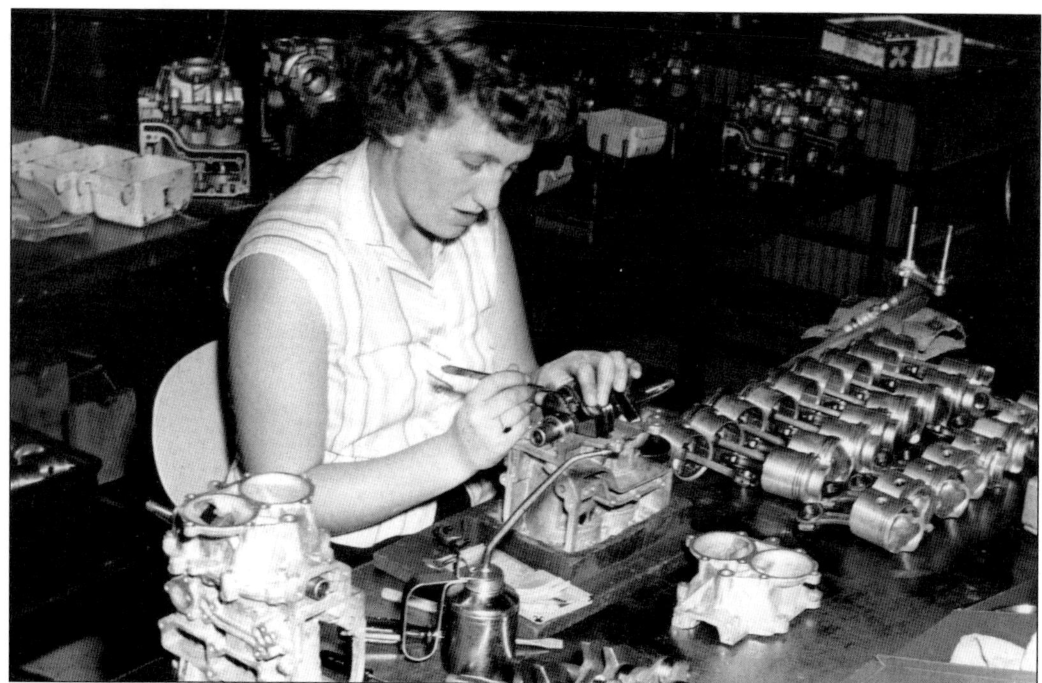

A worker assembles small-engine parts in the West Bend of Canada's Barrie, Ontario, plant in October 1960. The Barrie plant focused on manufacturing outboard motors, whose many intricate parts were assembled by hand.

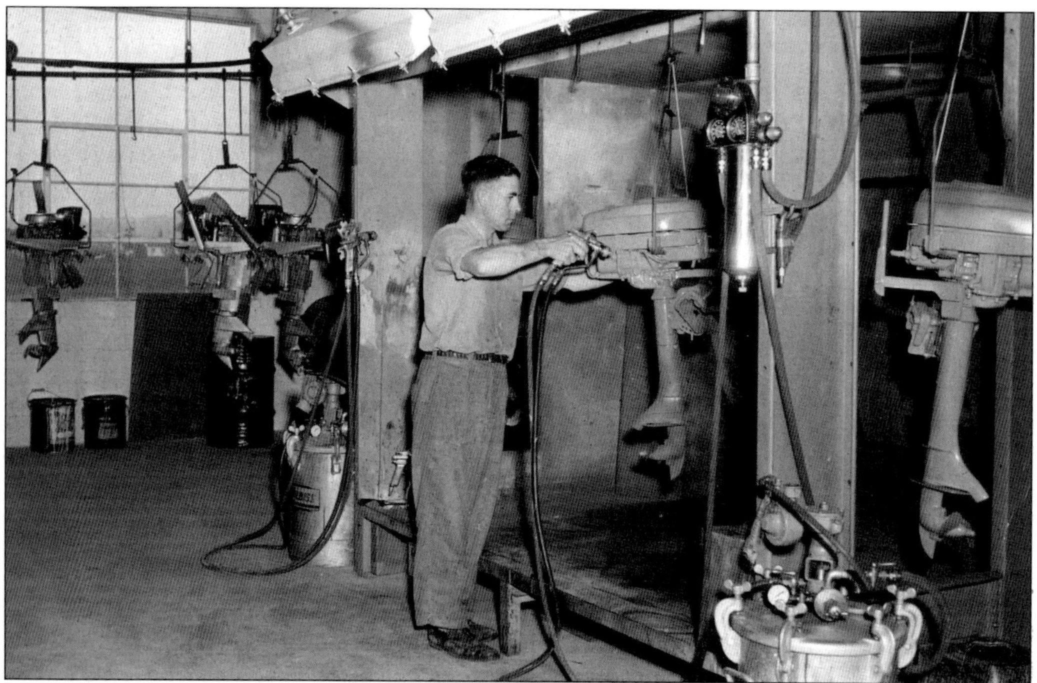

The success of the small-engine production at the West Bend Aluminum Company's Hartford plant led the company to begin operations in Barrie. Here the finishing touches are put on a motor by an employee at the Barrie plant.

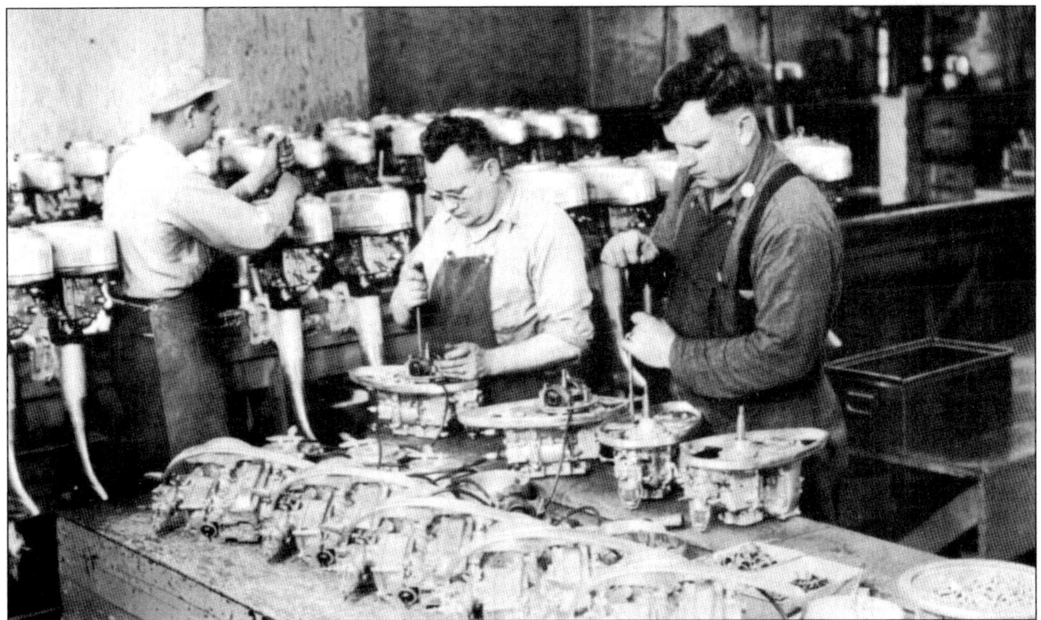

Workers assemble outboard motors at the West Bend Aluminum Company's Hartford plant around 1950. The company began the manufacture of outboard motors shortly after World War II. The first motor in production was the Elgin, which was sold directly though the Sears, Roebuck and Company. By 1948, two more motors were in production, one under the West Bend name and the second called the Shark.

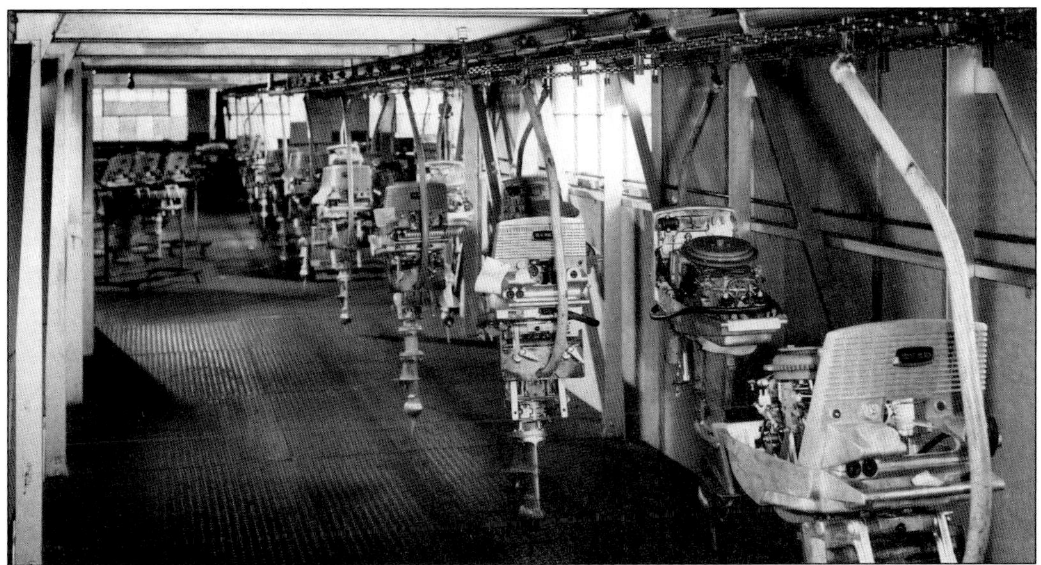

This scene shows the West Bend Company's Shark outboard motor under construction at the Hartford plant around 1961. The Shark outboard line was introduced in 1959. By 1961, West Bend was the manufacturer of the broadest range of horsepower models in the outboard industry. Motors ranged from 80 horsepower down to only 2 horsepower. The outboard motor portion of the company was sold to Chrysler Corporation in 1965 in order to focus on the cookware side of the business. The Barrie, Ontario, Canada, plant, which had been devoted to motor production, was converted to a distribution center.

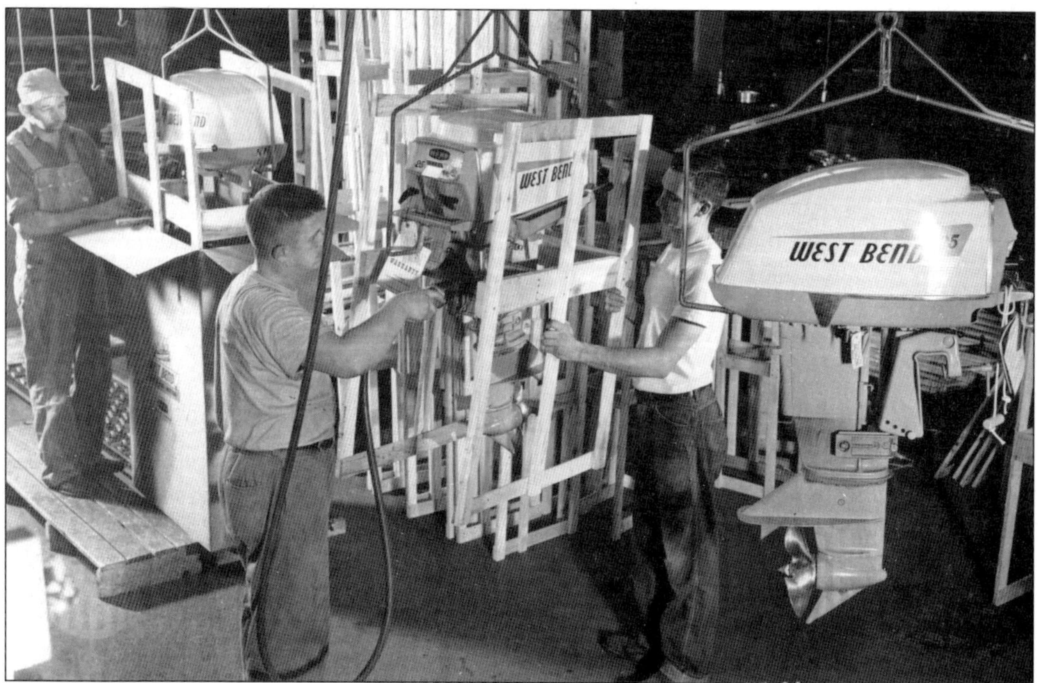

This photograph shows the end of the production line for the West Bend outboard motors around 1961. The last step before shipping the motors was packaging. Not as simple as cookware, the motors were encased in a wooden frame before being boxed.

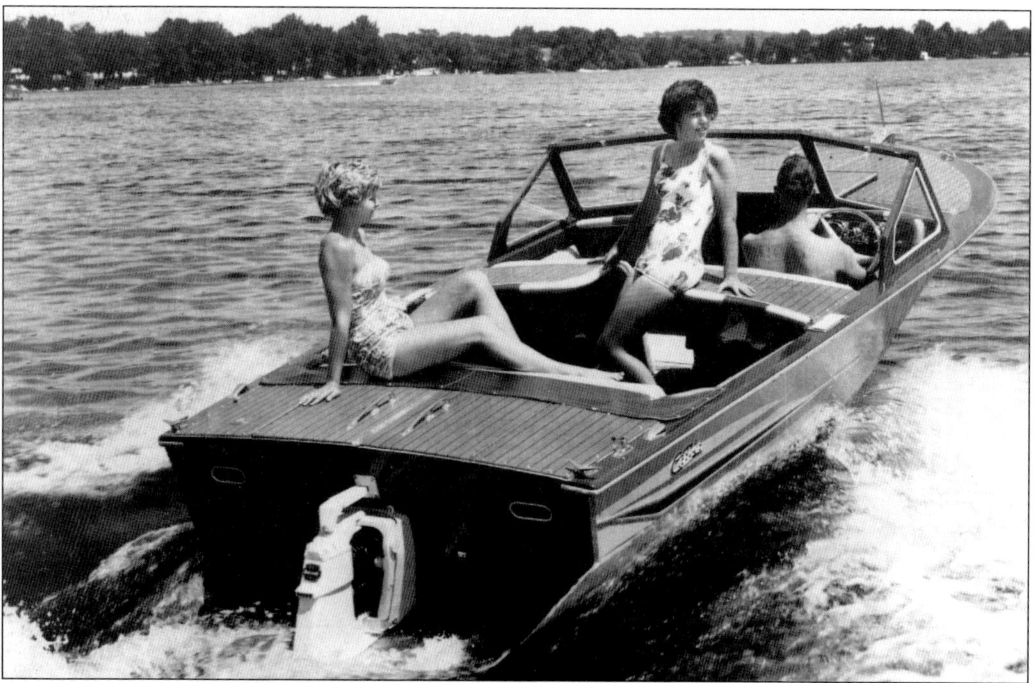

This is the eventual use of the motors produced at the West Bend Company Hartford and Barrie, Ontario, plants. This promotional photograph shows the leisurely lifestyle of owning a West Bend outboard motor.

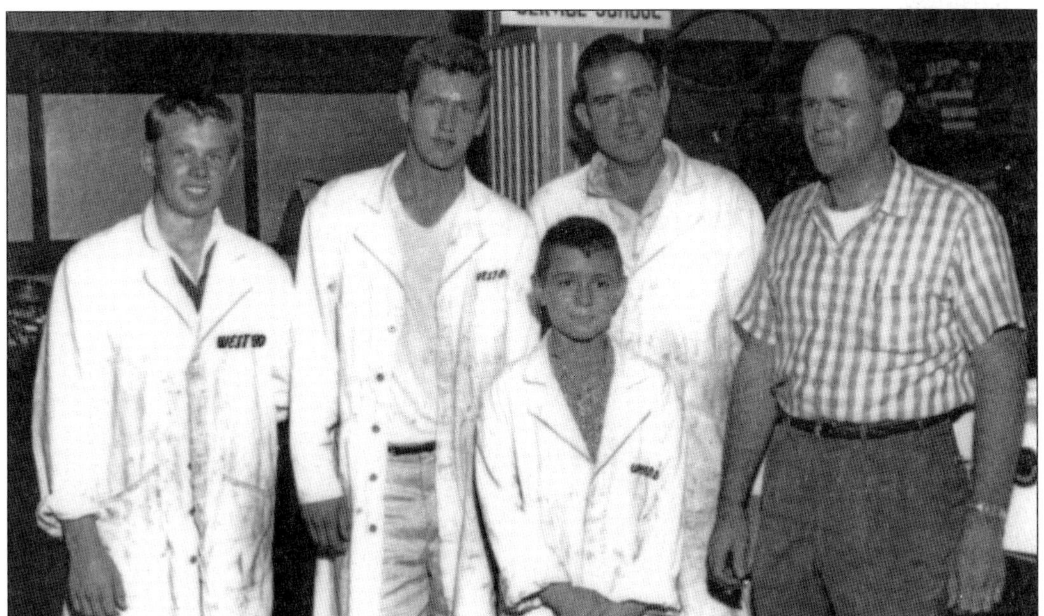

The West Bend Aluminum Company offered training sessions to dealers in the service and repair of outboard motors. In 1959, David Jacobson, on the left, and Dave Bie, second from the left, of Roseville attended the program. The other attendees are not identified. The program was established to train dealers to repair the products they sold. Upon completion of the training session, attendees obtained a certificate of completion.

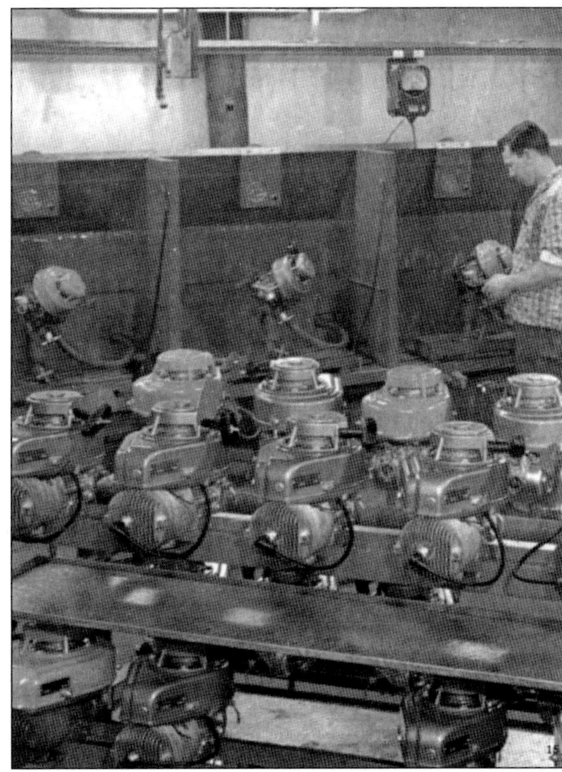

Power Bee engines wait to be inspected around 1961. Along with outboard motors, the West Bend Company produced small two-cycle engines at the Hartford factory. The Power Bee, first manufactured in 1956, was designed specifically for the sport of karting.

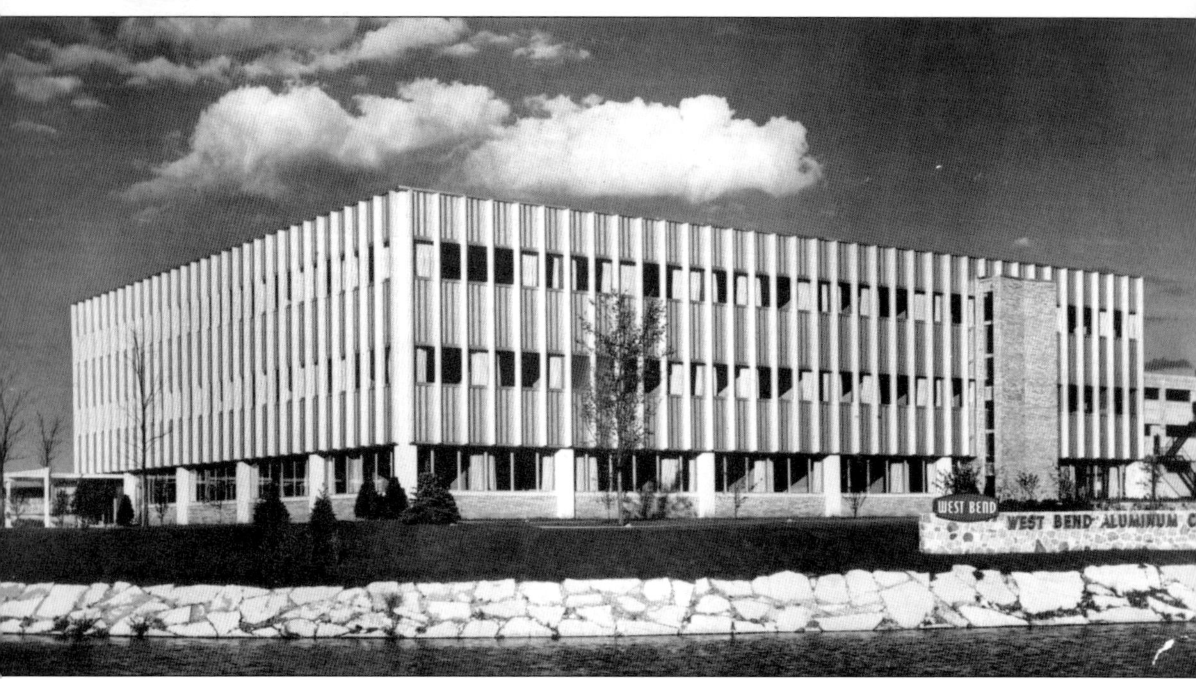

The new West Bend Aluminum Company office building was completed in 1958. The building was nicknamed the Golden Palace after the color of the facade as well as the lucrative business the company was doing.

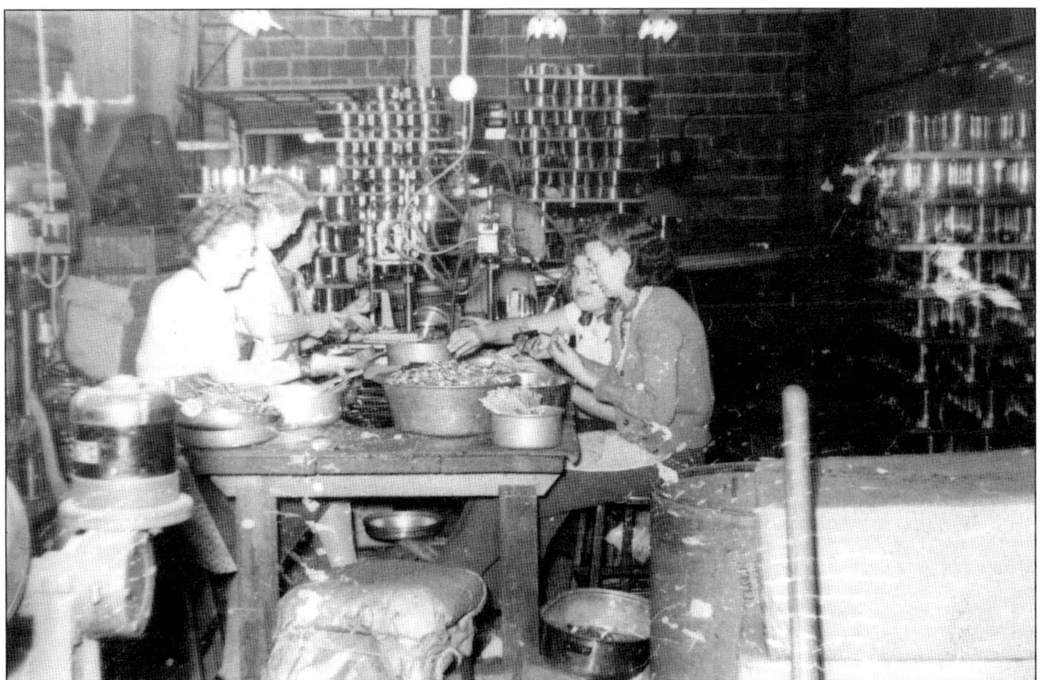

This is a very early photograph of the interior of the Kewaskum Utensil Company. It shows women sorting though handle components around 1947.

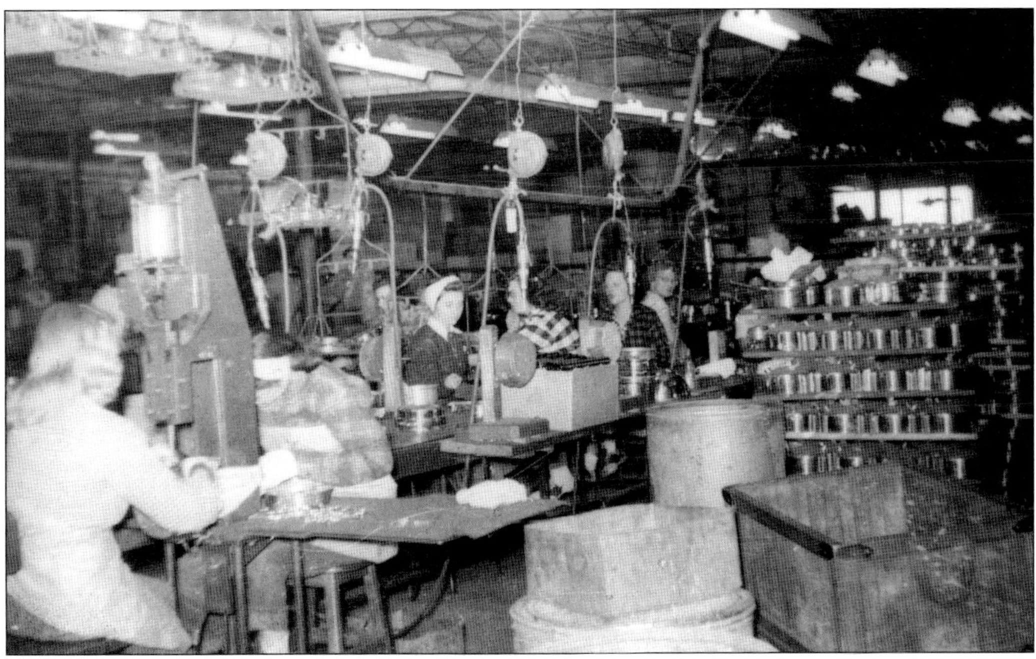

This is another very early photograph of the workings of the Kewaskum Utensil Company from around 1947. Here women are part of the production line. The cords hanging from the ceiling are automatic screwdrivers for attaching handles. Products can be seen on shelves in the background.

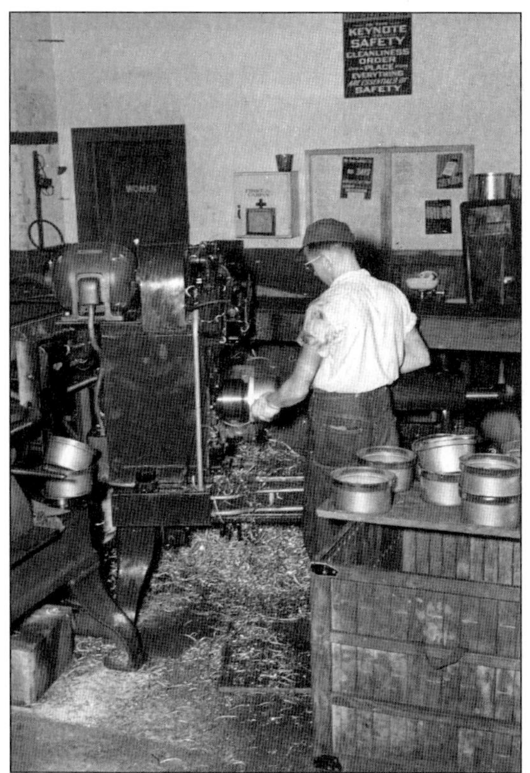

Here a Regal Ware Inc. worker trims a utensil around 1950. Trimming took place after the basic form of the utensil had been completed. The piece was placed on a form and turned, while the worker trimmed the edge of the utensil to make it even. Notice all of the trimmings on the floor.

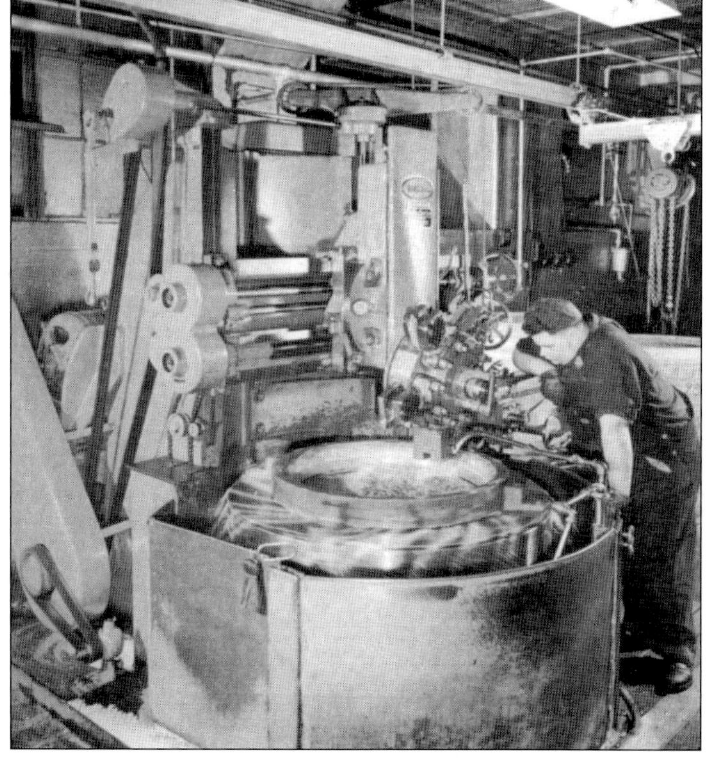

A Regal Ware Inc. worker looks on as a Bullard vertical boring mill does its job around 1950. The boring mill was one of several machines located in the machine shop and part of the Tool and Die Department. Components for the machines that did the product manufacturing were created, mended, and fine-tuned on the boring mill.

This photograph shows the Press Department at Regal Ware Inc. around 1950. Here the utensils took form. In the background, a visual inspection of finished product is taking place.

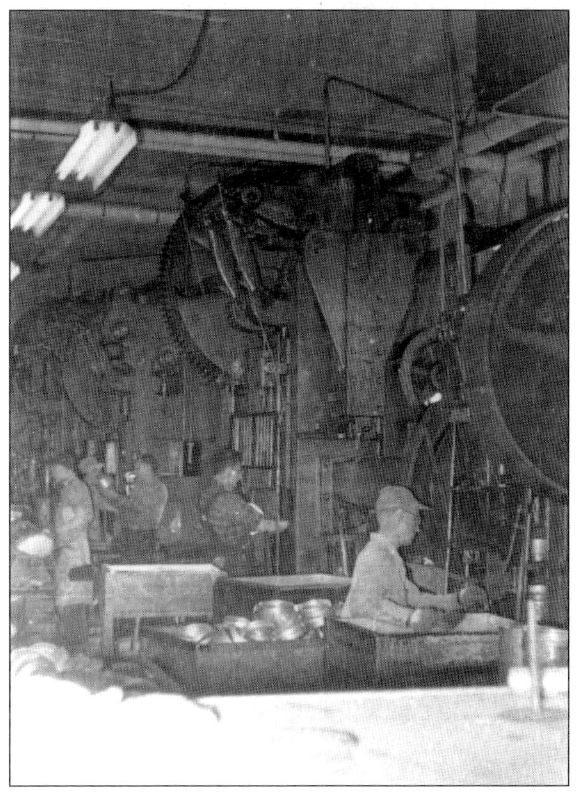

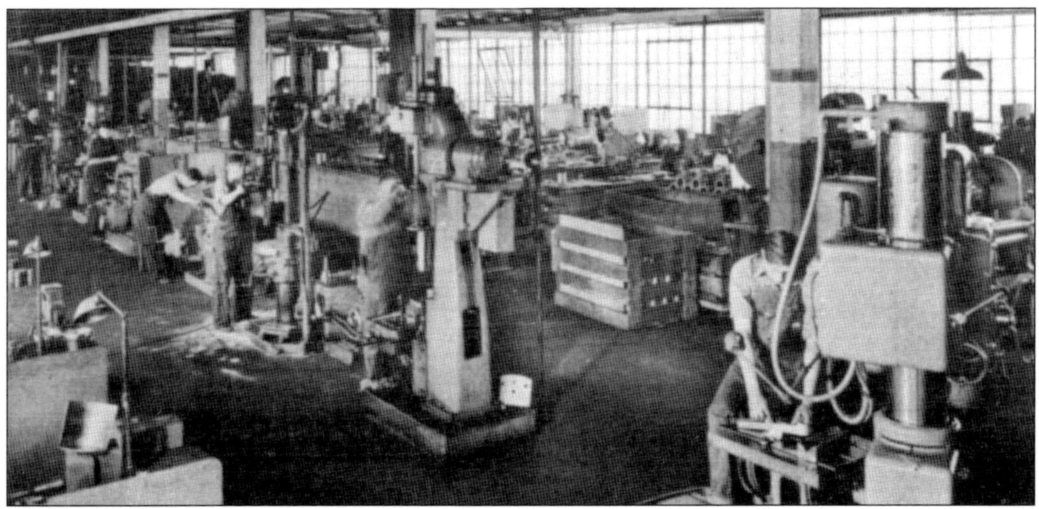

This view is of the West Bend Aluminum Company machine shop in about 1950. A large selection of machine tools allowed the company to not only produce its own product, but to do contract work for other manufactures as well. Tools in this room include radial drill presses, drilling machines, shapers, and horizontal and vertical mills. Part of the modern factory features included the large windows seen here. Windows allowed for a natural lighting condition, which saved on energy and was easier to work by.

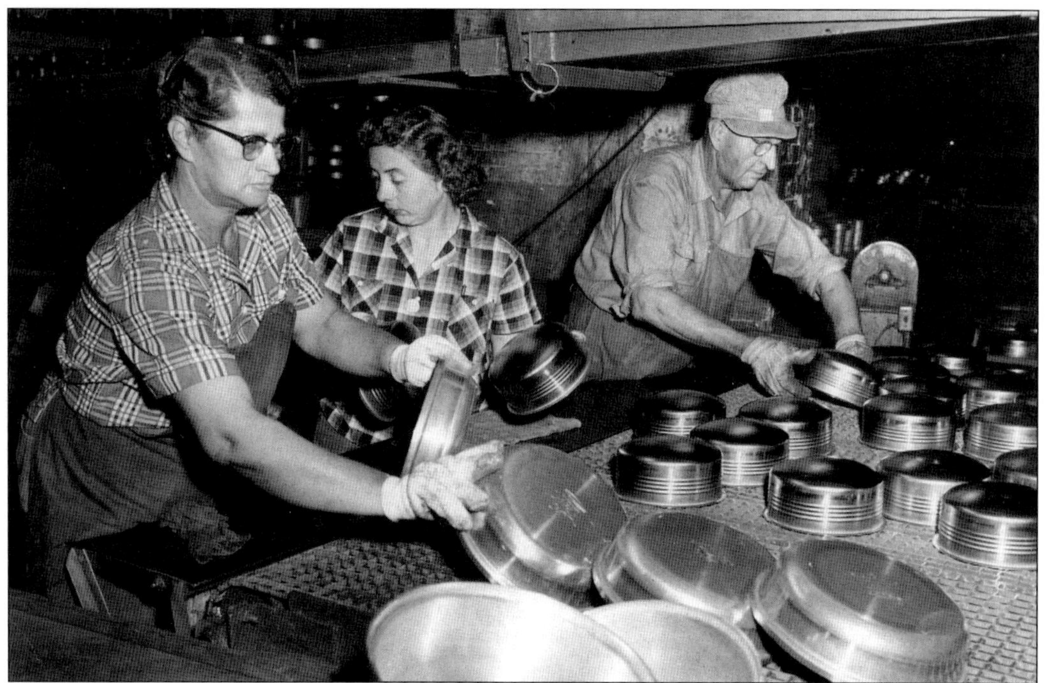

Regal Ware Inc. line workers are seen around 1953. The pans are fresh from the washer, which removed the residue from the drawing process. It was after the wash that the pans received their first major inspection.

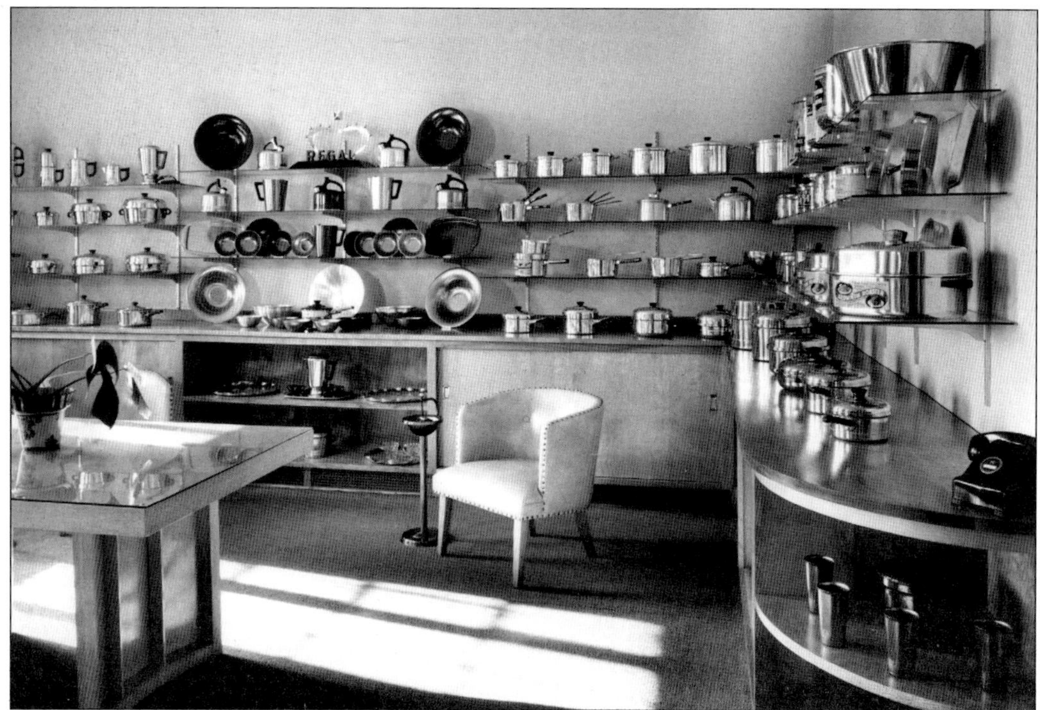

This display graced the Regal Ware Inc. office building in 1951. The display allowed visitors and salesmen to see and compare all of the company's products in one location.

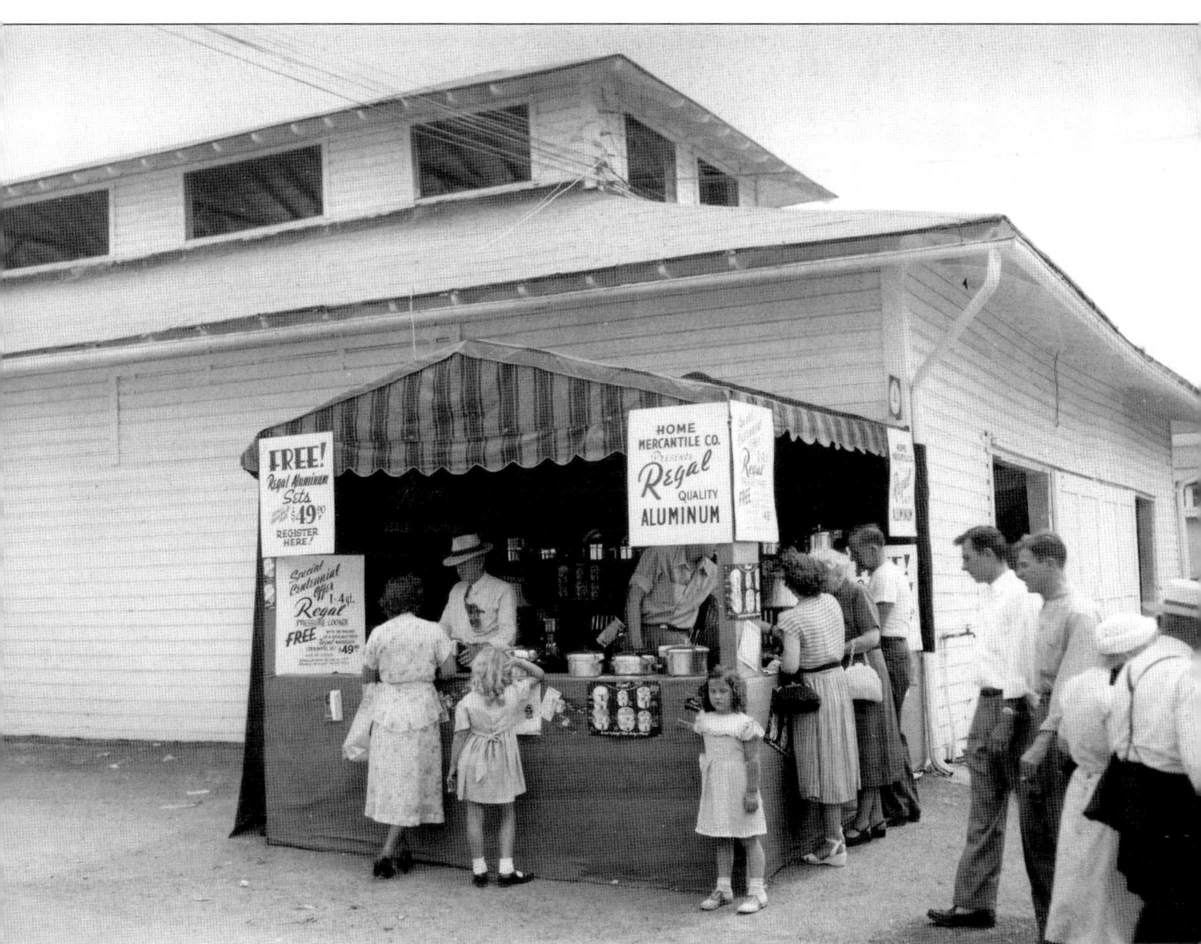

This photograph shows the Kewaskum Utensil Company's booth at the Wisconsin State Fair in West Allis around 1946. The company's trademarked product line, Regal Aluminum, can be seen advertised on the signs. It was this name that the company capitalized on in 1951, when it was adopted as the company's name of Regal Ware Inc.

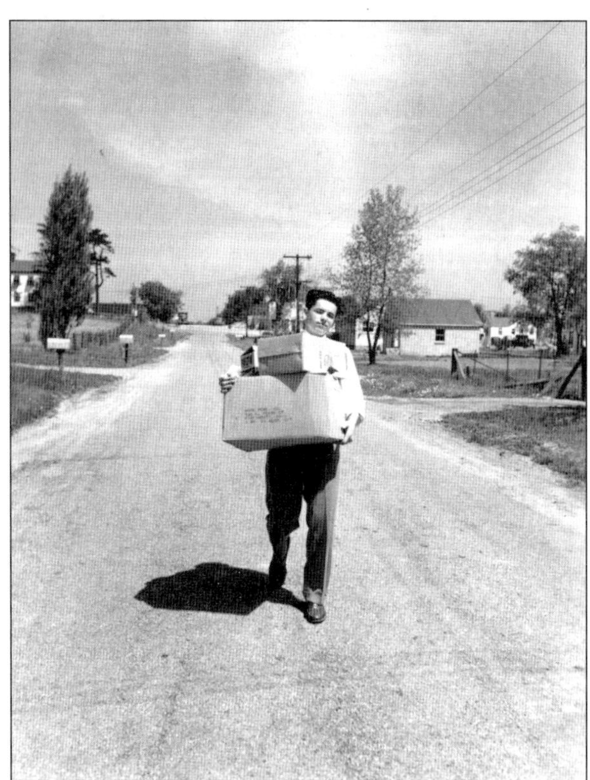

A young direct salesman is walking his beat around 1950. Many of Regal Ware Inc.'s early products were sold though the direct or door-to-door method.

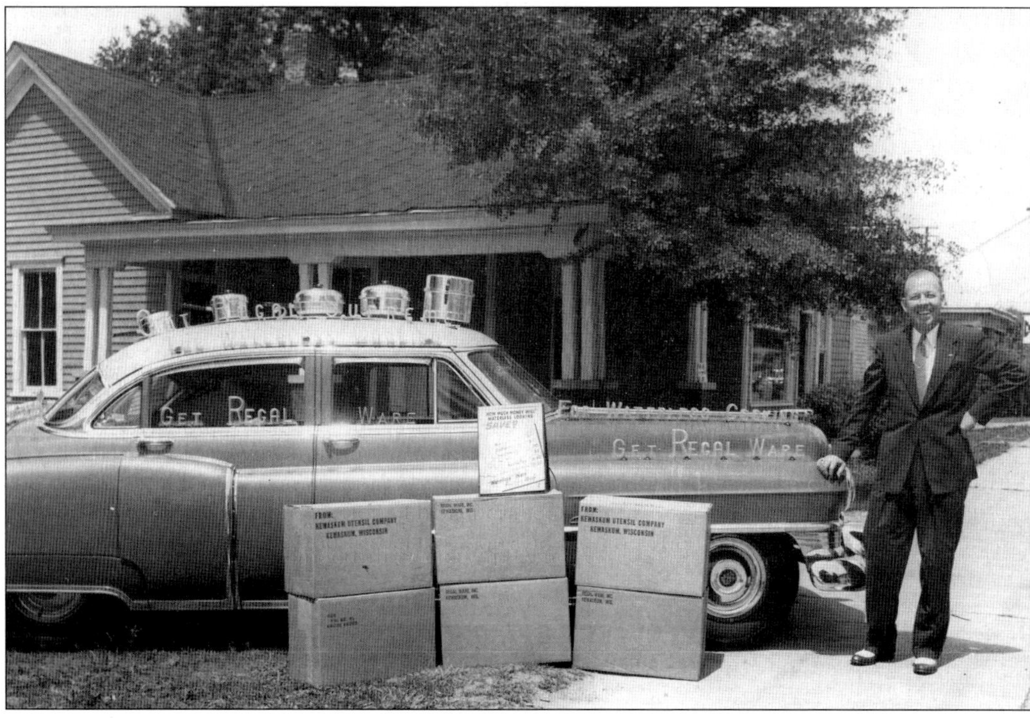

George Van L. Green poses by his sales car. Green's territory was Atlanta, Georgia, and he became the first salesman to sell $1 million of Regal Ware merchandise in one year, around 1948.

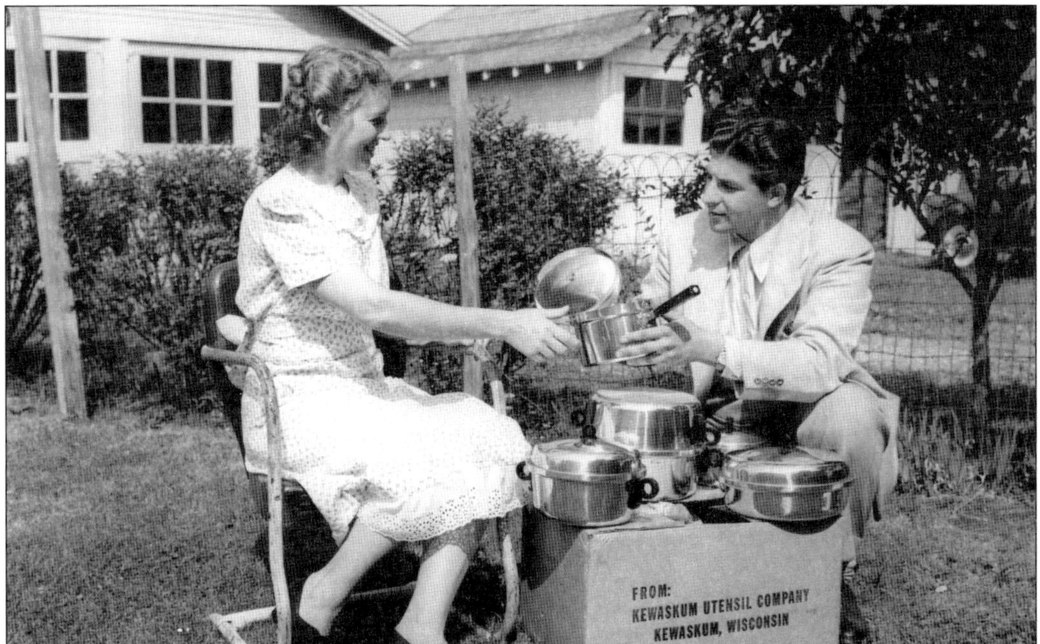

A Kewaskum Utensil Company salesman presents Mrs. I. W. Gates of Waukegan, Illinois, with her set of Regal Ware cookware. Mrs. Gates was a contest winner and this photograph appeared in the Waukegan *Sun and News* newspaper. The distinctive loop-style handles were a design change proposed by Edna Oster, J. O. Reigle's assistant.

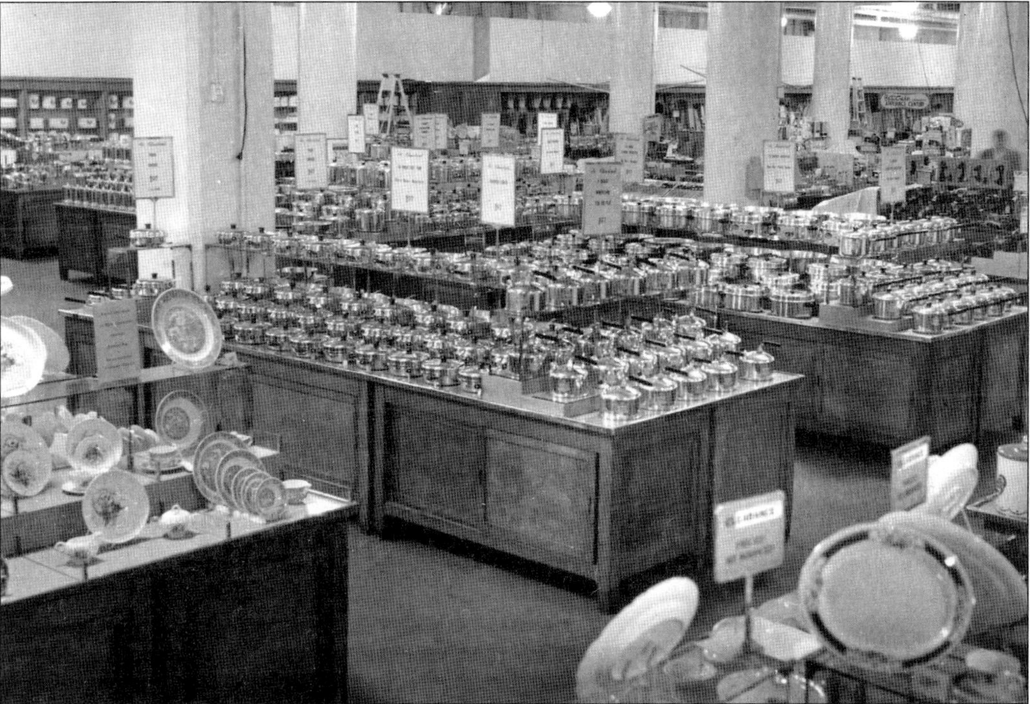

Regal Ware products are on display in February 1950. This sale at the Montgomery Ward store in St. Paul shows the extent of Kewaskum Utensil Company's products.

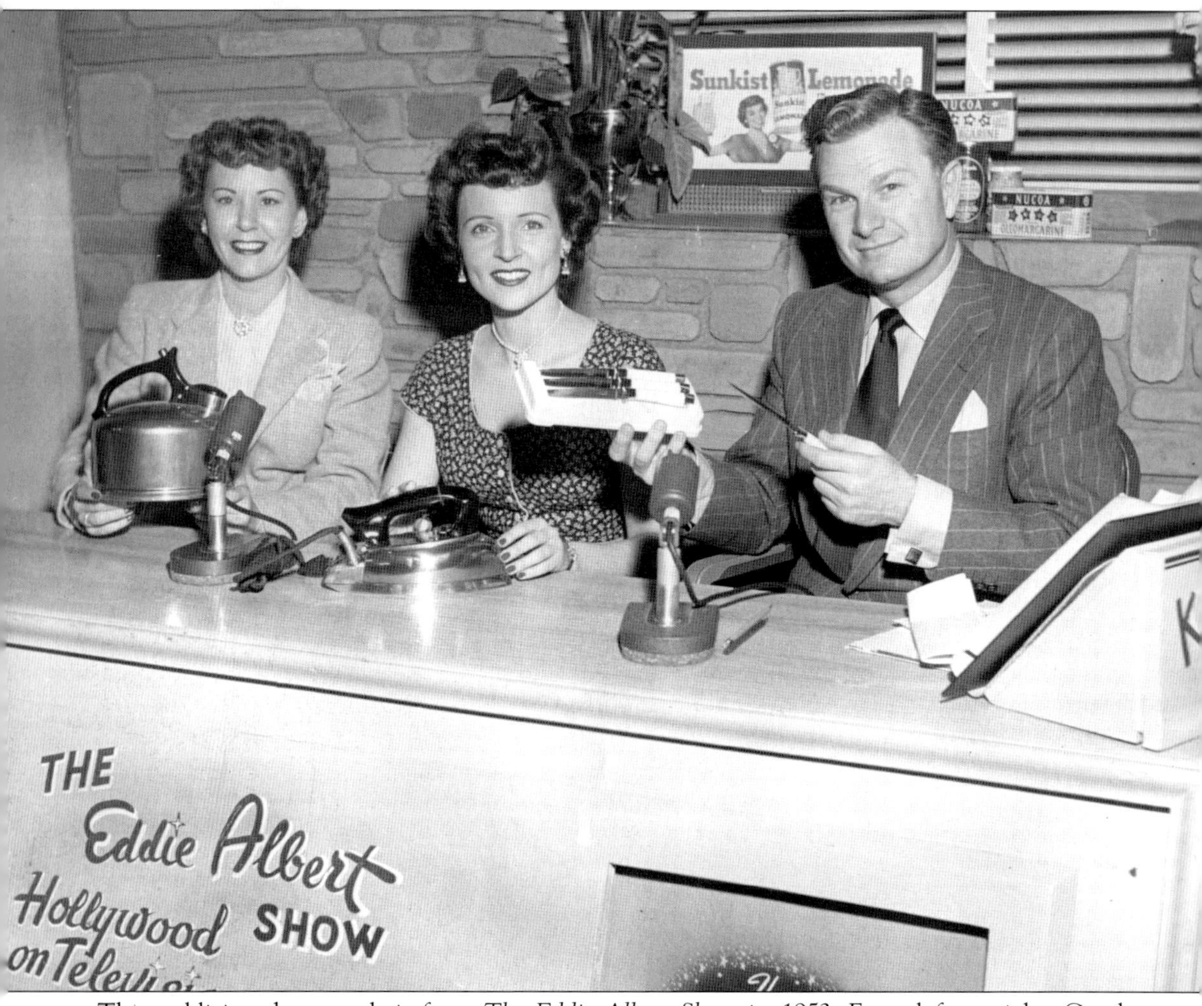

This publicity photograph is from *The Eddie Albert Show* in 1953. From left to right, Carol DeNise, Betty White, and host Eddie Albert hold Regal Ware products. The CBS show ran for only one year, but Regal Ware was able to showcase its wares on the program.

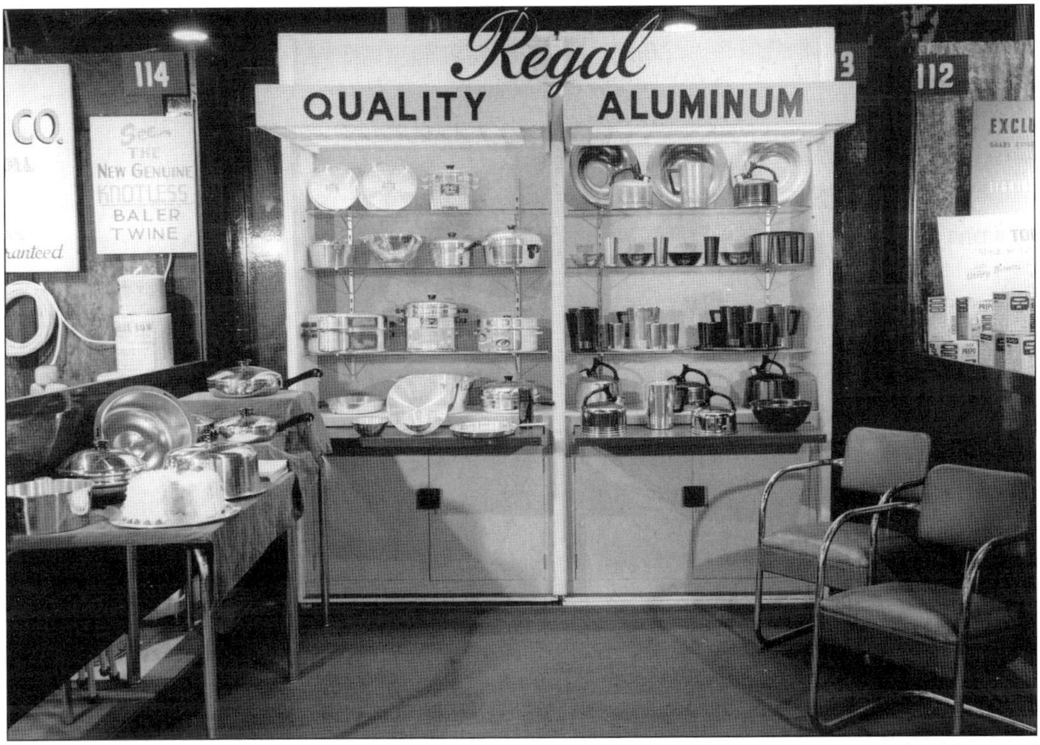

This photograph shows one of the earliest exhibits from Regal Ware Inc. at the National Housewares Manufacturers Association annual housewares show in Chicago. Regal Ware Inc. first appeared at the show shortly after its inception in 1946. The show was sponsored by the National Housewares Manufactures and began in 1939.

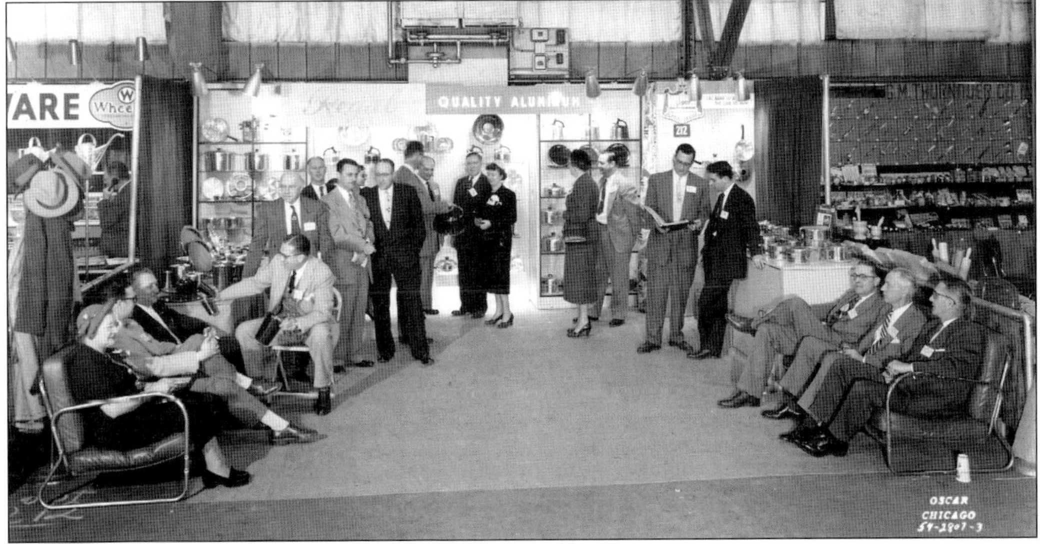

The National Housewares Manufacturers Association show in Chicago was the most important trade show in the industry. More contacts could be made in one day at the show than in several days of phone calls. This was Regal Ware Inc.'s exhibit at the Chicago Cookware Show in 1954.

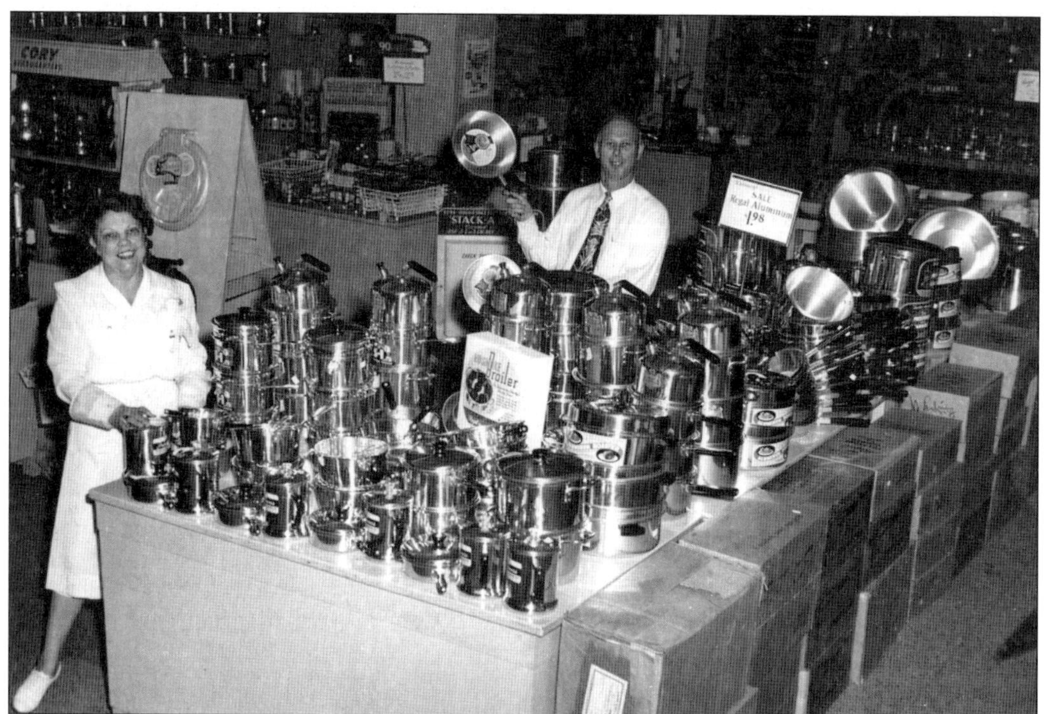

Workers at a store named Whitney's in Albany, New York, proudly show off the Regal Ware display around 1950.

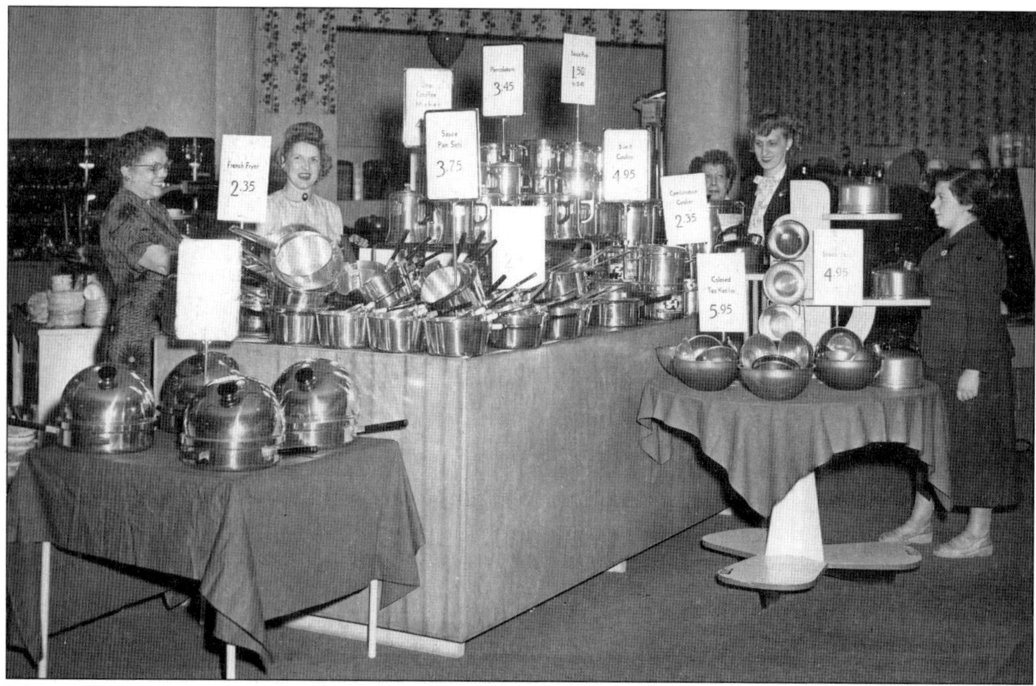

This unknown promotion exhibit, probably in New York, is just one example of the range of products manufactured by Regal Ware Inc. in the 1950s. The only person identified is Norine Bonham, standing on the far left. Bonham was a department store demonstrator for Regal Ware Inc.

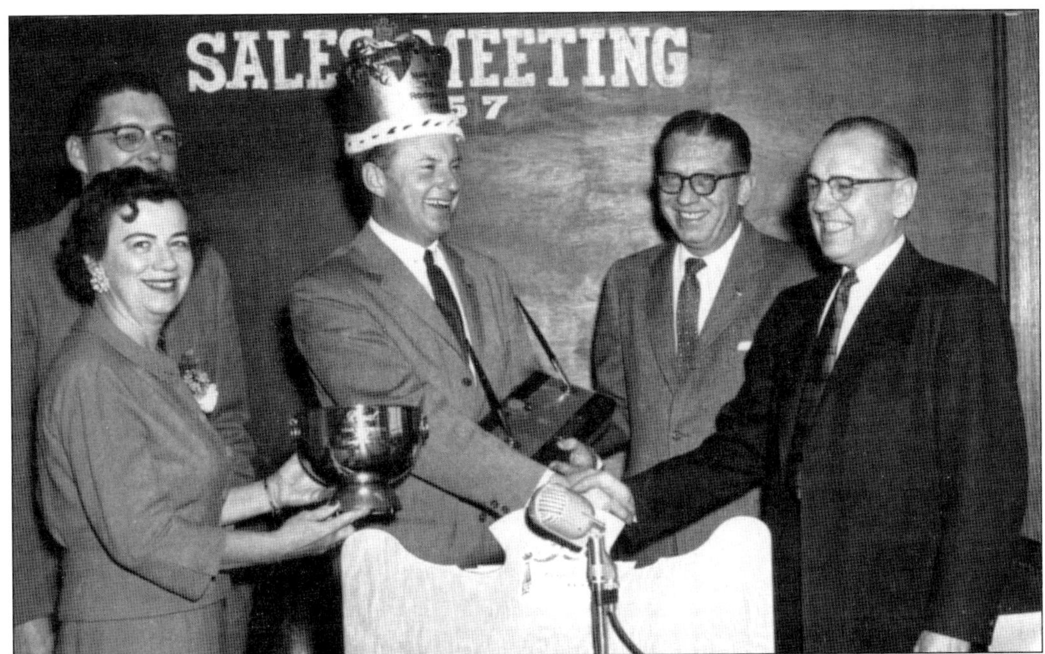

In December 1957, Rodger Dunston was crowned Man of the Year at the annual Regal Ware Inc. sales meeting. Crowing a Man of the Year introduced friendly competition between the salesmen, giving incentive to increase sales. To the left of Dunston is Edna Oster and Jim Reigle. Shaking his hand is J. O. Reigle. Standing next to J. O. is Red Freaman, vice president of sales.

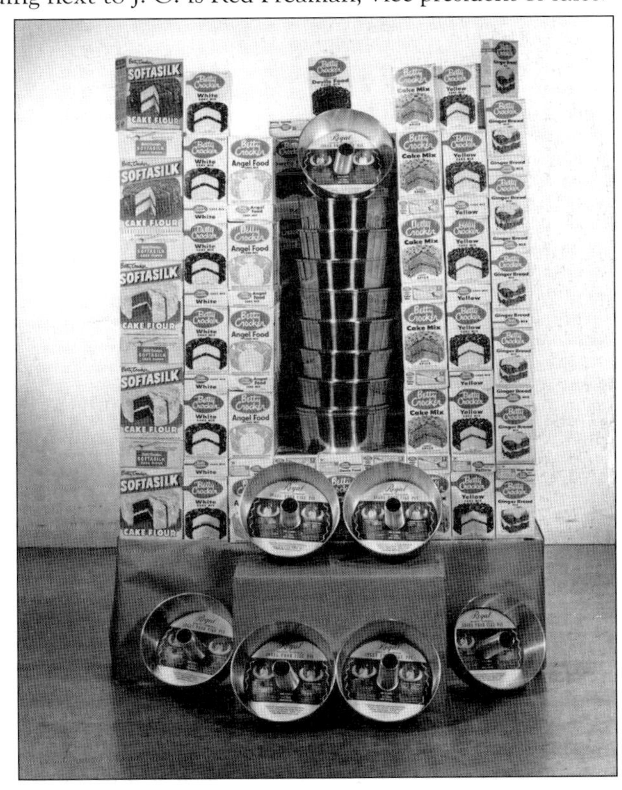

One popular marketing arrangement that many companies participated in was that of premiums. Here Regal Ware Inc. teamed up with Betty Crocker to create a display with cake pans and mixes. The partnership with Betty Crocker continued in 1976 when Regal Ware Inc. and General Mills's Betty Crocker entered a licensing agreement that allowed a Betty Crocker approval ribbon to be placed on product packaging. The Regal Ware Inc. cookware that received the Betty Crocker endorsement was a premium cast aluminum line called Harvest and produced in the Wooster, Ohio, plant.

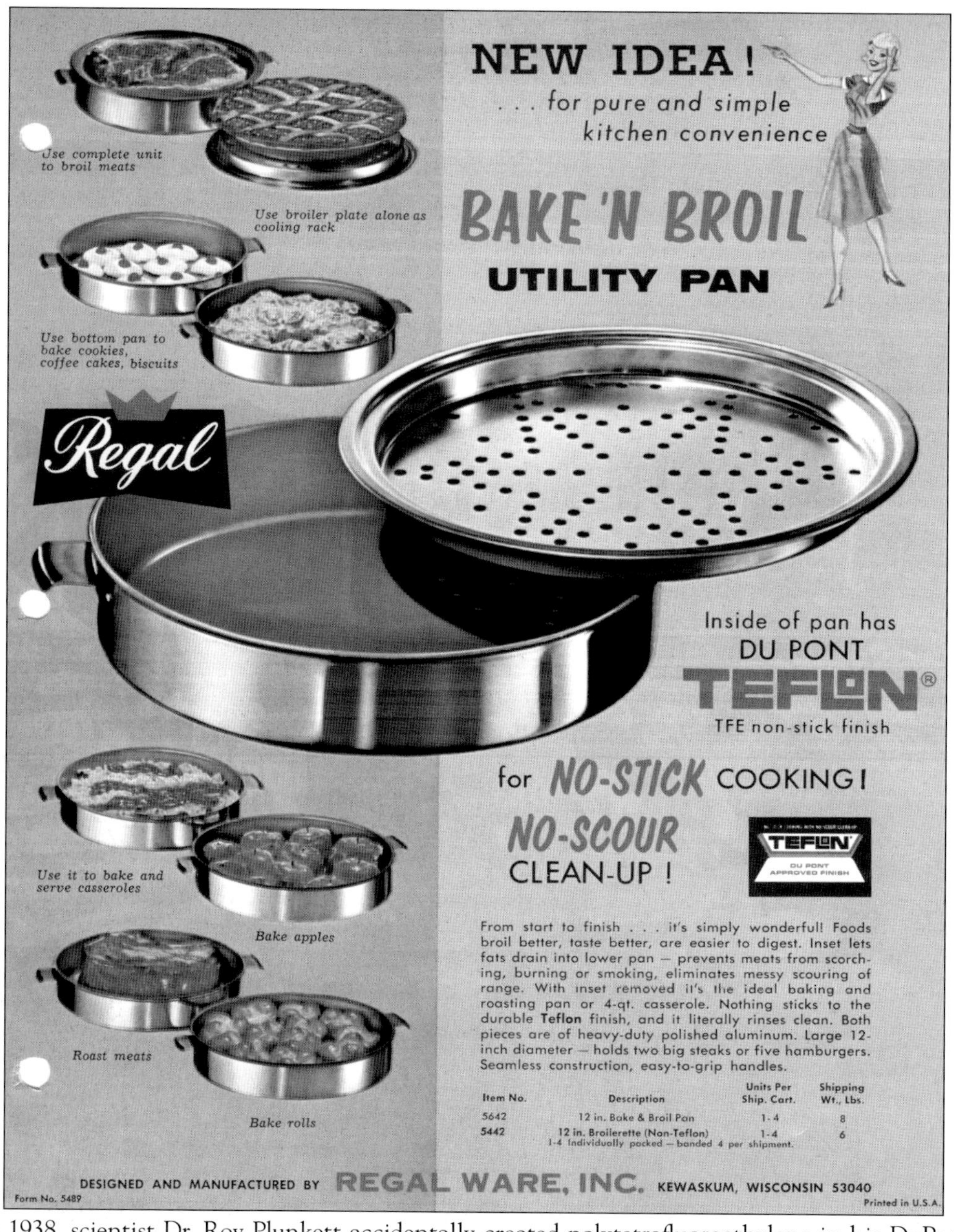

In 1938, scientist Dr. Roy Plunkett accidentally created polytetrafluoroethylene in his DuPont lab while researching refrigerants. DuPont trademarked the product in 1945 as Teflon. In 1954, Marc Gregoire discovered a way to allow the Teflon to adhere to aluminum. Gregoire began to make and sell Teflon coated aluminum cookware from his home. The business was so successful that it quickly became a large commercial enterprise. Soon Teflon was being used on cookware throughout the industry. The nonstick Teflon coating allowed easy clean up of cookware and healthier cooking as fewer fats had to be used to prevent sticking. Over the years several different types of nonstick coatings have been used on cookware, from Teflon II to Silverstone and Gem Coat. This Regal Ware Inc. advertisement from the early 1960s shows one of the many products utilizing the coating.

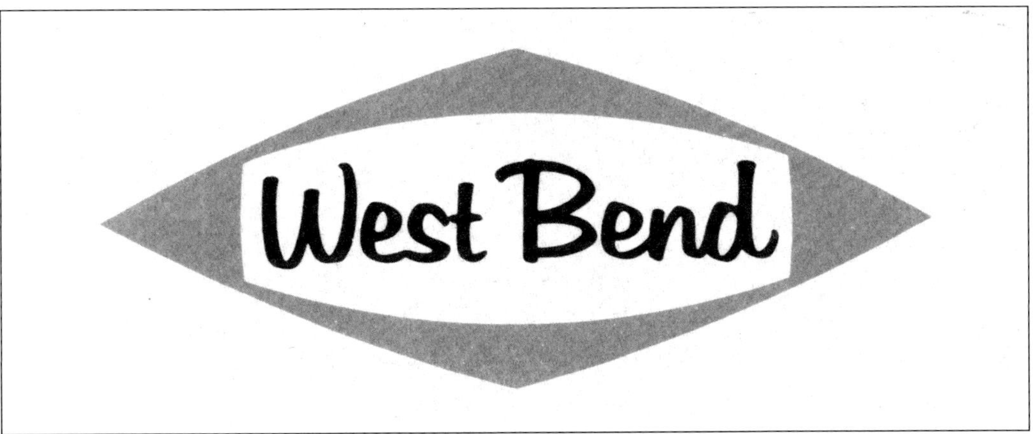

In 1961, the company's 50th anniversary year, the West Bend Aluminum Company dropped *Aluminum* from its name. Due in part to the diversification of products and materials used to produce products, the word aluminum was seen as limiting. This image shows the first logo used under the new name, West Bend Company.

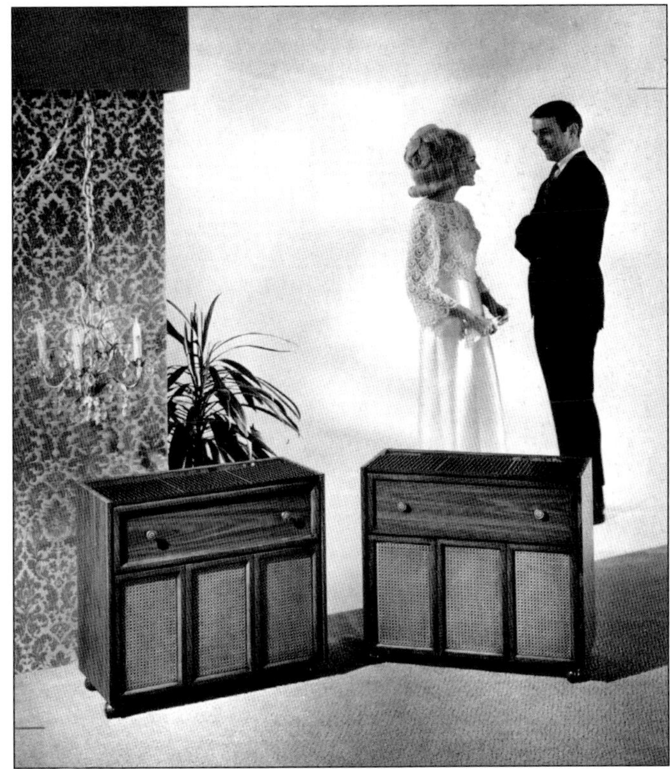

The Supreme Solid State Automatic Humidifier can be seen in this promotional photograph from the early 1970s. The design of the floor and table top humidifiers was made to look like a piece of furniture. Wood grain and styling helped the product blend in with the modern decor.

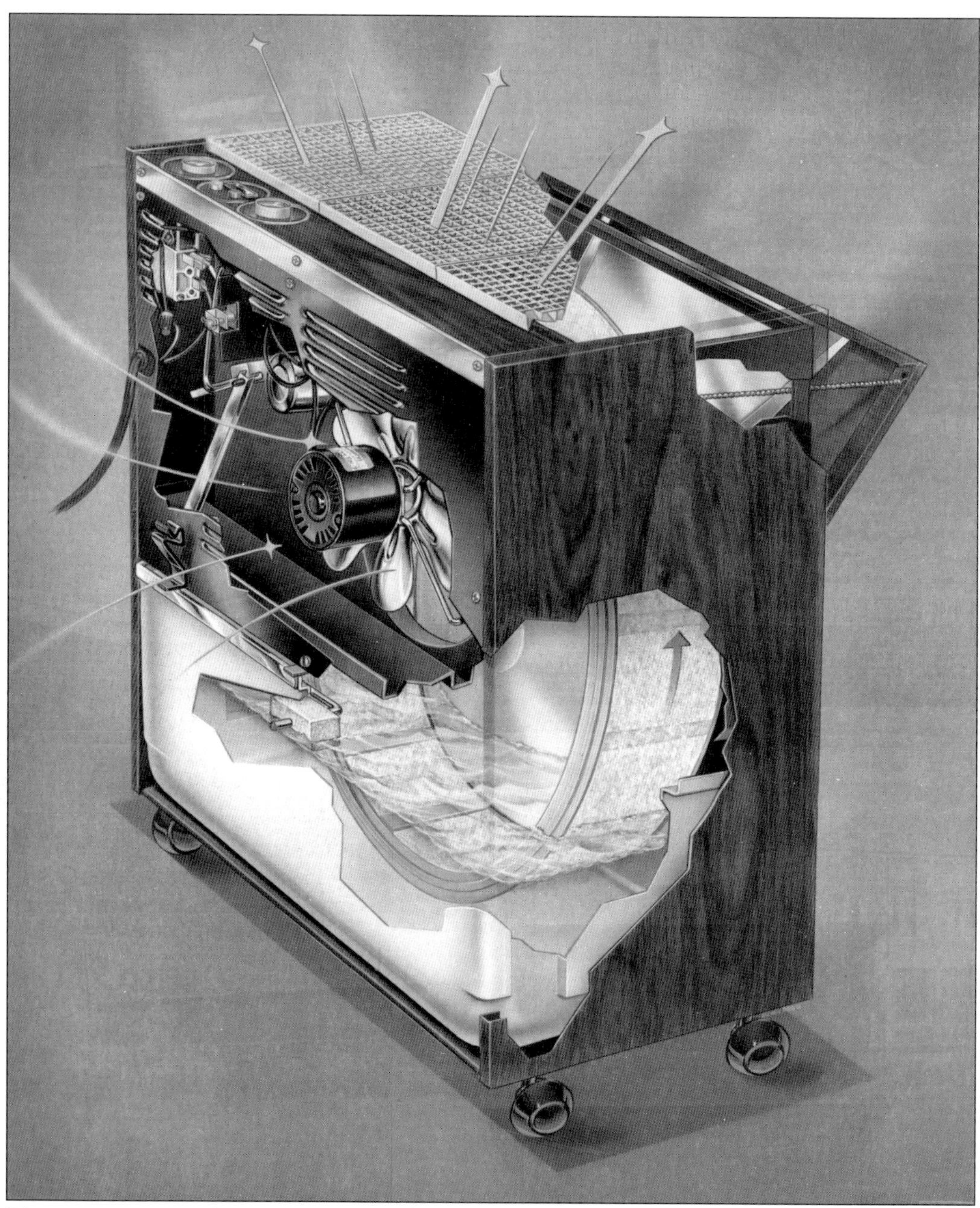

This cut away rendition shows the inner workings of one of the West Bend Company's humidifiers. Introduced in 1965, the humidifier line became one of the best-selling products of the company. Humidifiers were made under a separate home comfort department. This images shows the "water wheel" design, which helped to make the West Bend Company the world's leading manufacturer of portable humidifiers by 1972.

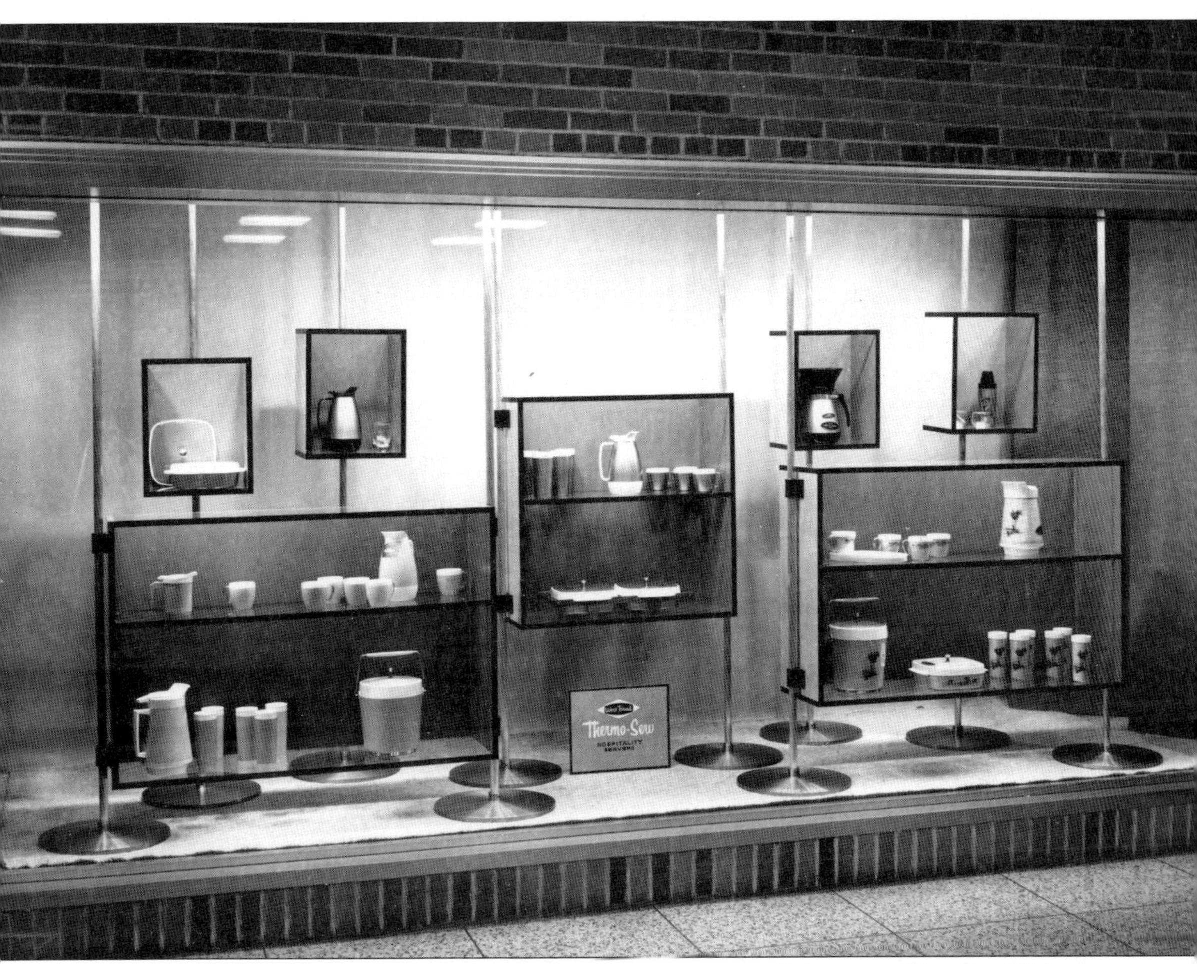

In 1965, the West Bend Company acquired NFC Engineering Company of Anoka. With this acquisition came a new market, that of insulated plastic serving ware, sold under the name Thermo-Serv. This photograph from around 1965 shows several of the products produced under the Thermo-Serv name. One of the most popular Thermo-Serv products was the coffee server, seen in the second box from the left in this photograph. Its distinctive copper-colored body with black lid, handle, and base could be seen in countless diners and restaurants across the country.

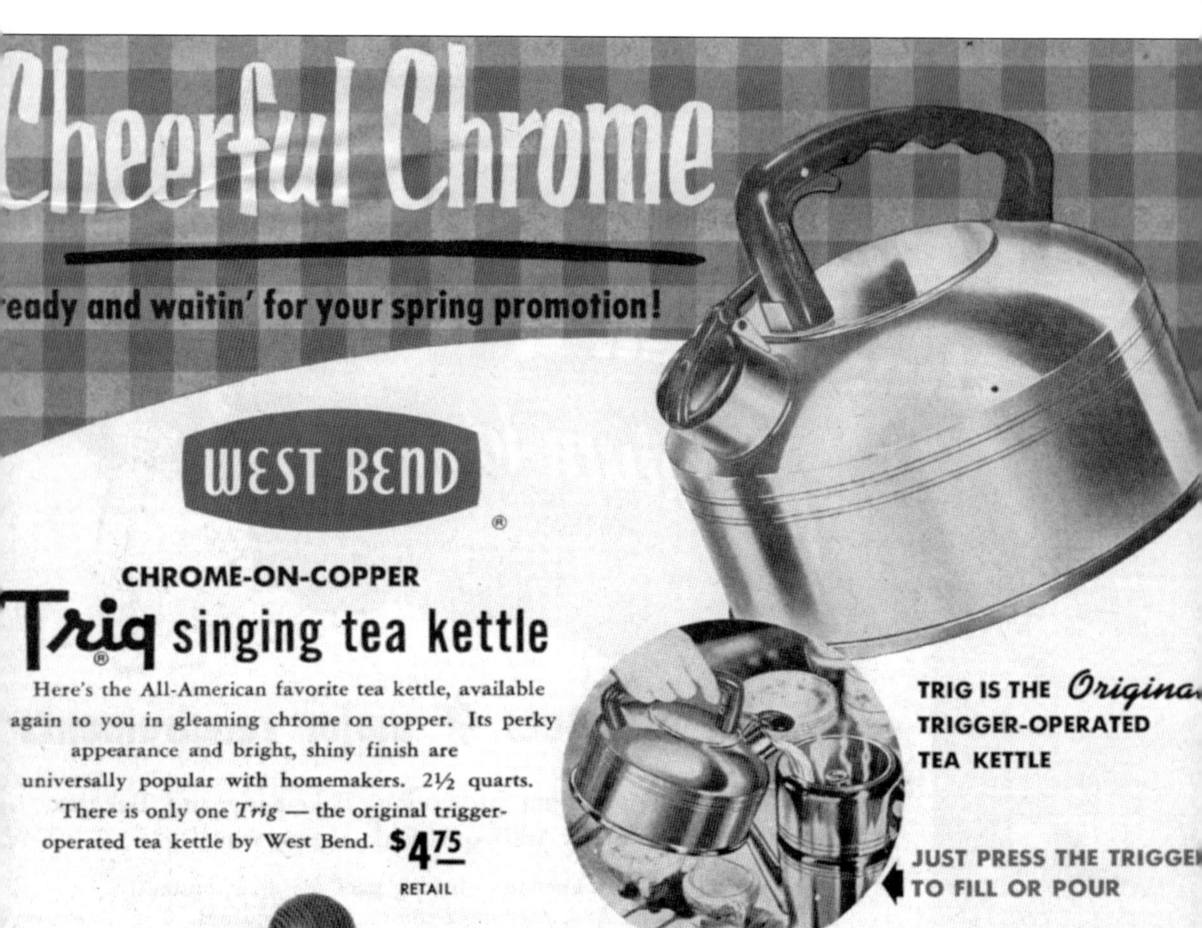

One of the most popular items in the West Bend Company product line, the Trig singing teakettle was first produced in the early 1950s. The first Trig teakettles were made completely of copper and were four quarts in capacity. A redesign made the pot smaller, two and a half quarts, and stainless steel with a copper bottom. The move was made to stainless steel to reduce the amount of cleaning needed to the outside of the pot. The copper bottom was added to improve heat conduction. By 1968, production was at nearly 50,000 per month, with 15 million total made by the fall of 1968. The kettle could be opened with one finger pulling a trigger, allowing the other hand free to turn on the water to fill the kettle.

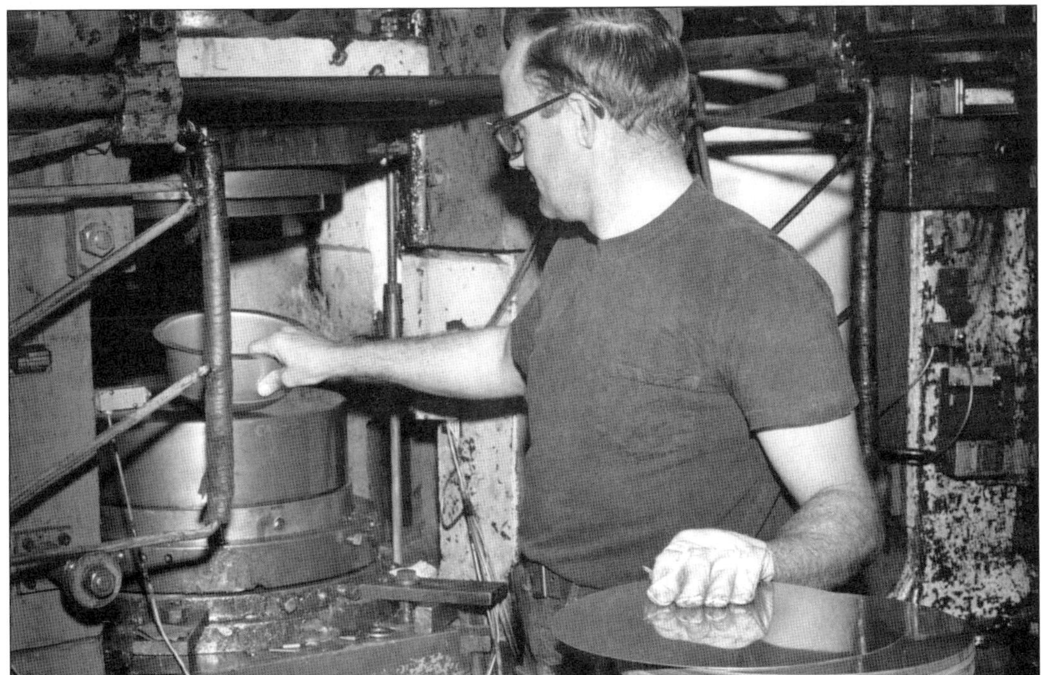

The first step in creating the West Bend Company's Trig teakettle takes place in the press room. Here Sylvester Peters places blanks in a press to draw the tea kettle's shell. The Trig was one of the most popular items made by the company.

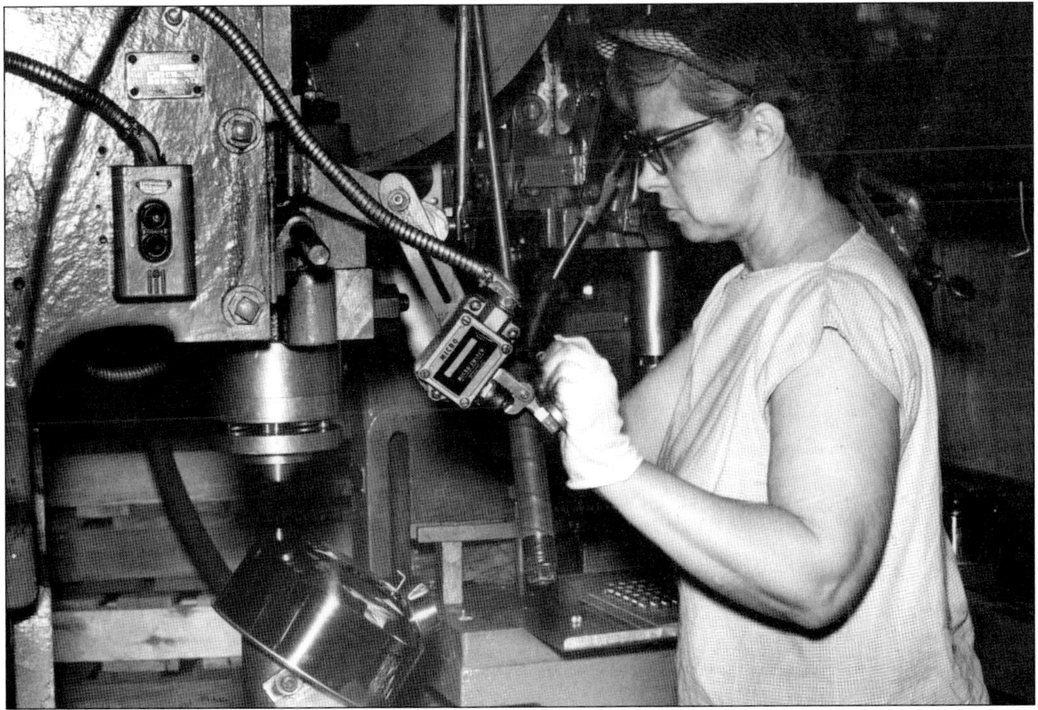

In another step in the production of the Trig teakettle, Elma Thom welds on the handle connections.

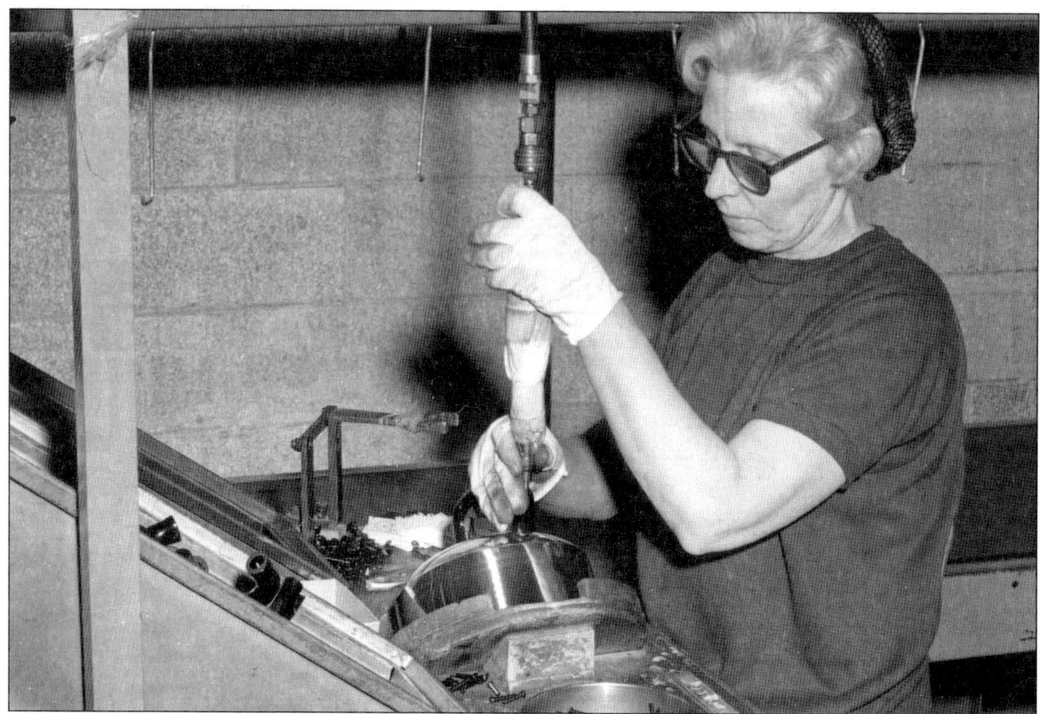

After the Trig's handle connections were welded on, the handles could be attached. In this photograph, Sara Rauscher attaches the handles.

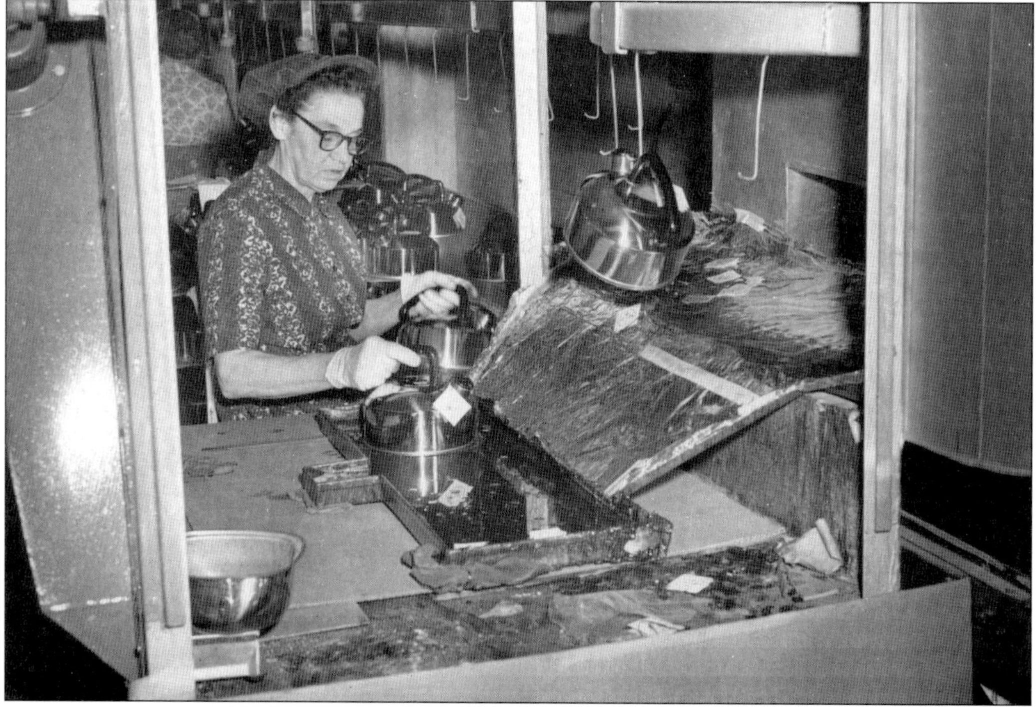

This photograph shows a step in the production of the West Bend Company's Trig teakettle in 1968. Here Nora Wickman lacquers the copper bottom of the pot before it is hung to dry.

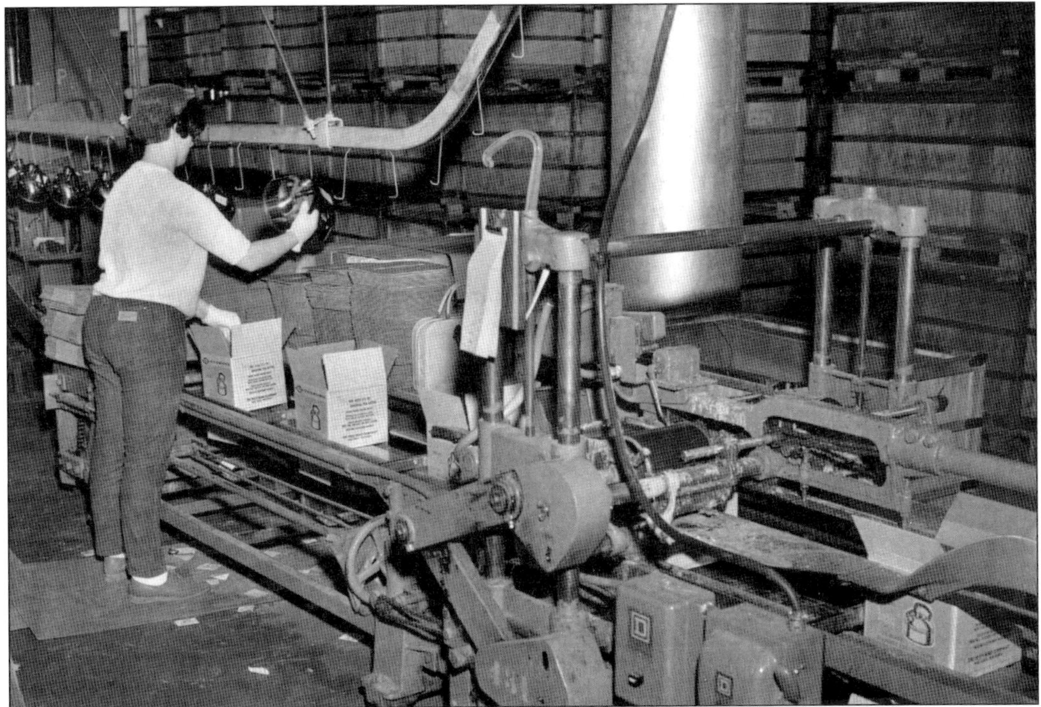

The final step in Trig teakettle production is the inspection and packing. Here Janice Berger does that at the end of the line.

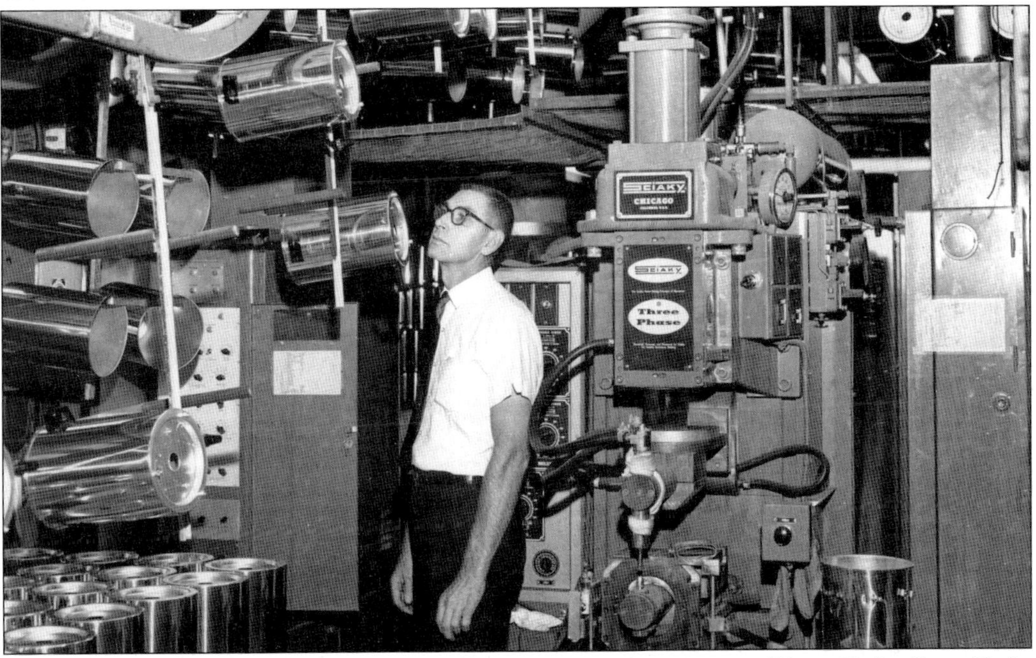

Don Puls, a West Bend Company industrial engineer, inspects a run of Party Perks. Party Perks were large-capacity coffeemakers that were a hugely successful seller for the company after they were introduced in the 1950s. Designs and colors of the Party Perk adjusted to the changing fashions. The first models were aluminum, like those seen here.

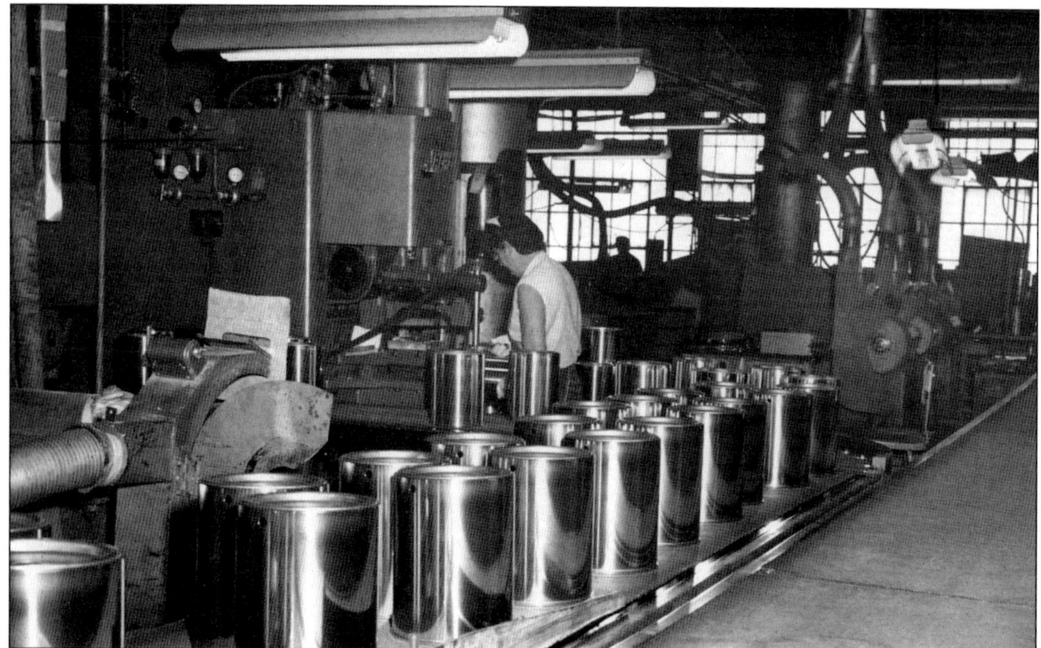

The Party Perk coffee percolator was one of the West Bend Company's largest sellers. The successor to the large drip coffee urns, the Party Perk was introduced in the 1950s. Nearing finish, this photograph shows the Party Perk shells moving toward Leona Monday. She will weld on handle connections.

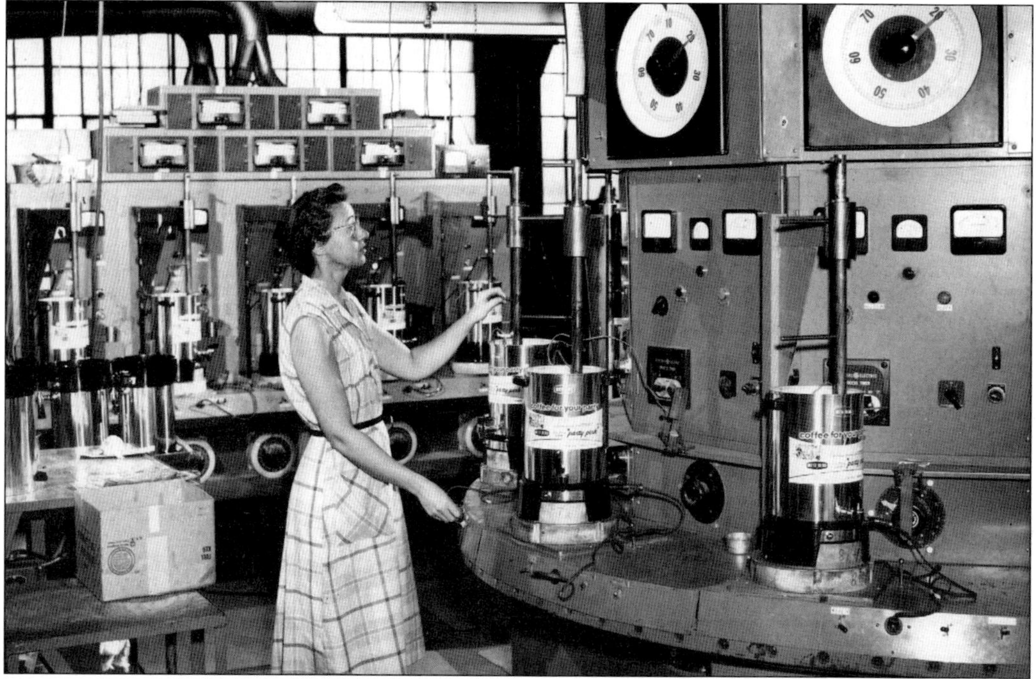

A West Bend Company employee does final quality control tests on Party Perk coffeemakers in 1961. This revolving test panel was designed by West Bend Company engineers to make testing more efficient.

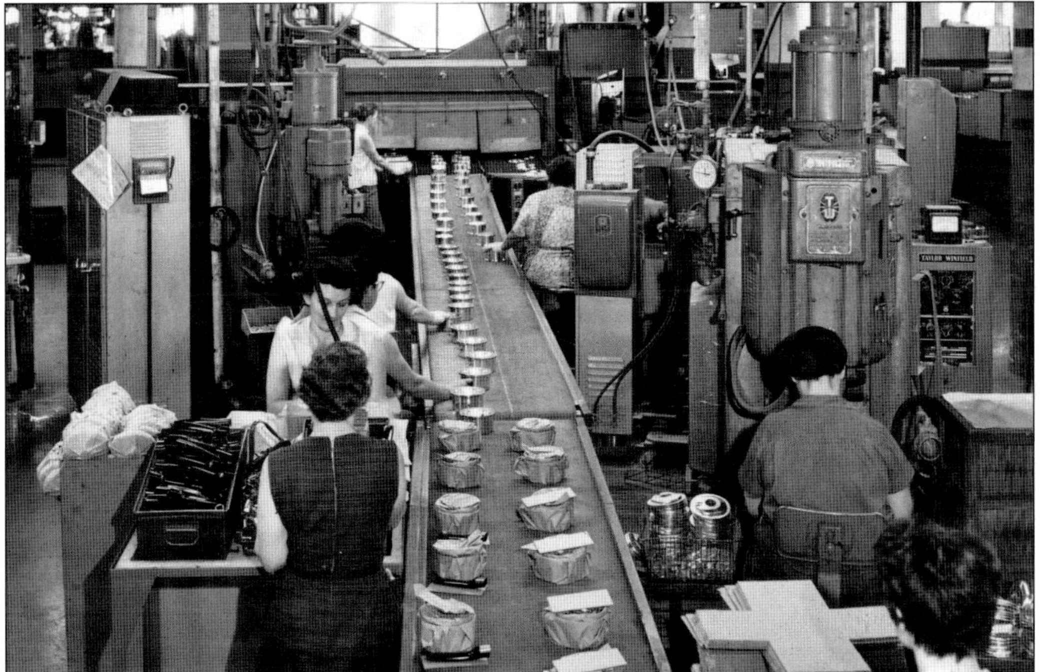

This photograph shows the West Bend Company's packing line in 1961. After a final wash and dry, the pans were hand wrapped and a handle added before being packaged. The pieces seen in this photograph were from one of the West Bend Company's waterless cookware lines.

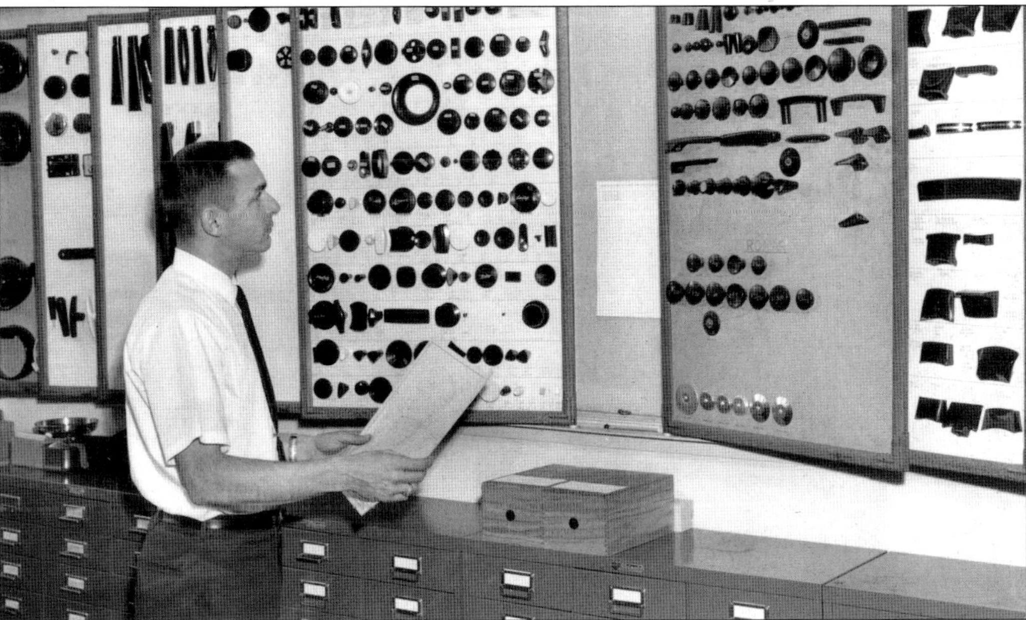

Ron Babros scrutinizes the phenolic supply boards. Babros was a development engineer at the West Bend Company. This department was responsible for new design concepts, visual and functional sample creations, as well as new part and component improvements for existing products. In 1967, the year this photograph was taken, the department brought 42 projects to completion.

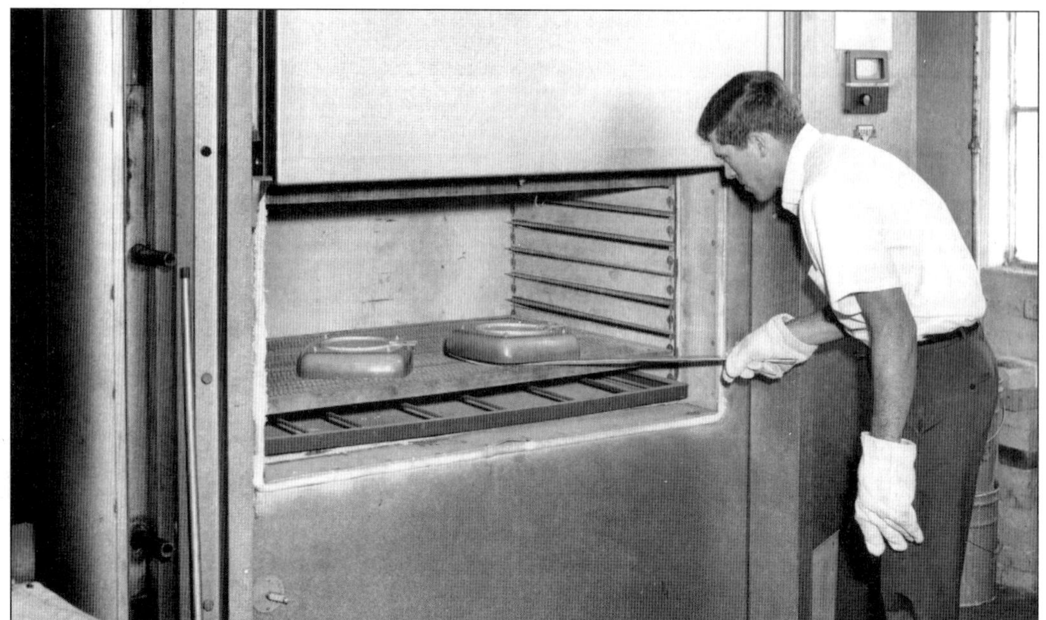

Lee Luehring, an engineer working in the Development Engineering Department, fires a porcelain coated pan in a high heat oven. The West Bend Company's Development Engineering Department initiated new designs. The department created mock ups of concepts for initial feedback and testing. It also monitored competitor's developments and designed improvements for existing products.

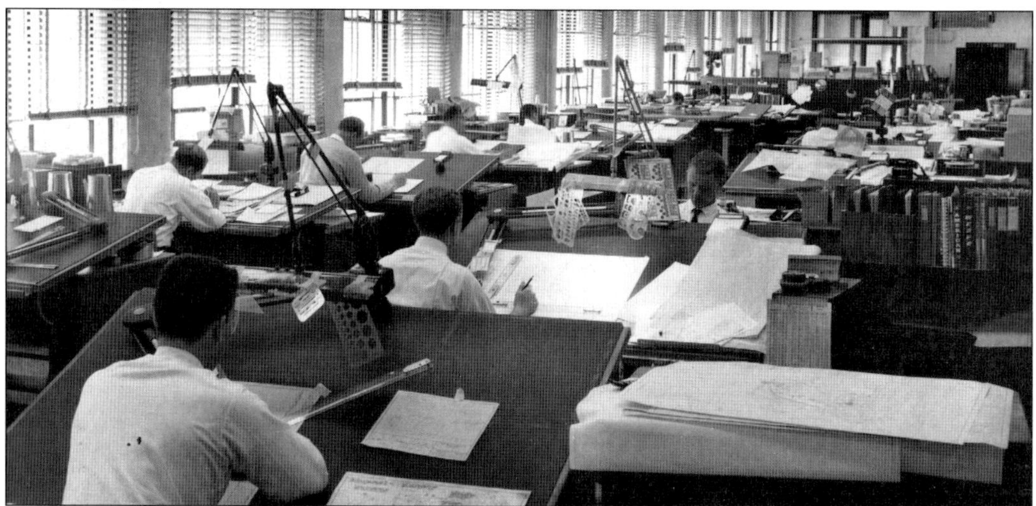

This photograph shows the Production Engineering Department in 1967. The company had four different engineering departments. Beginning with designs generated by the Development Engineering Department, Production Engineering determined the sequence of operations needed to create the product. They also designed and maintained the tooling equipment involved in production and developed the techniques and methods for the utilization of the equipment they designed. The department was divided into four divisions. The tool and die design and product engineering division employed 30 men in 1967. Another division was charged with machine design, while the third was engaged in electrical engineering. The final division was in charge of cost estimation.

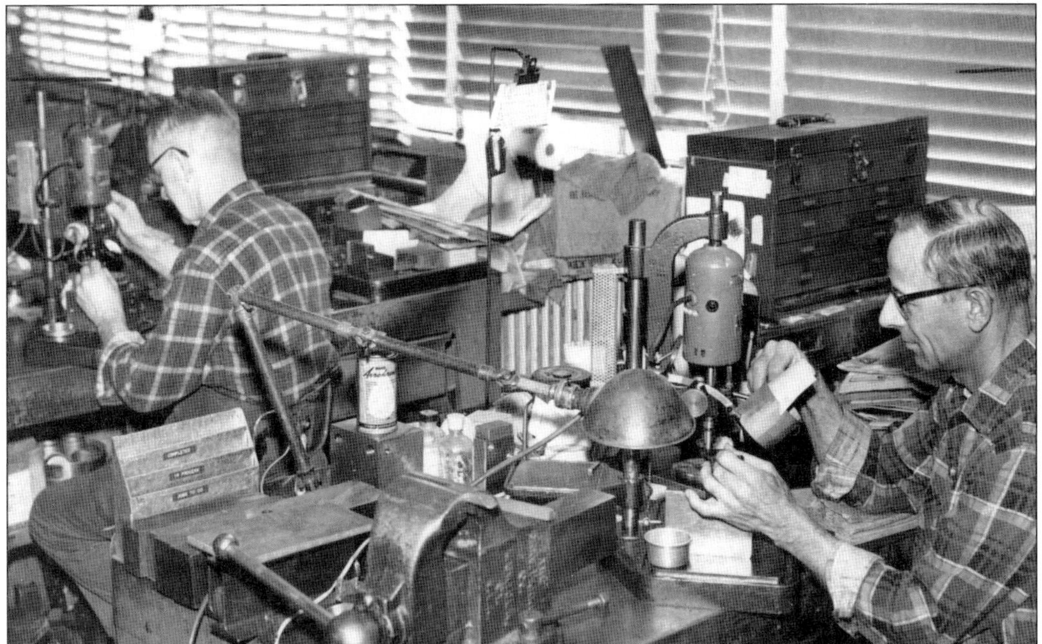

William Utech (left) and Ed Schacht make new dies for percolator baskets, the part that holds the coffee grounds, in the West Bend Company's Tool and Die Department in 1967. At the time, the Tool and Die Department had 47 employees. The department was tasked with creating dies for production from engineering drawings as well as keeping used dies in working order.

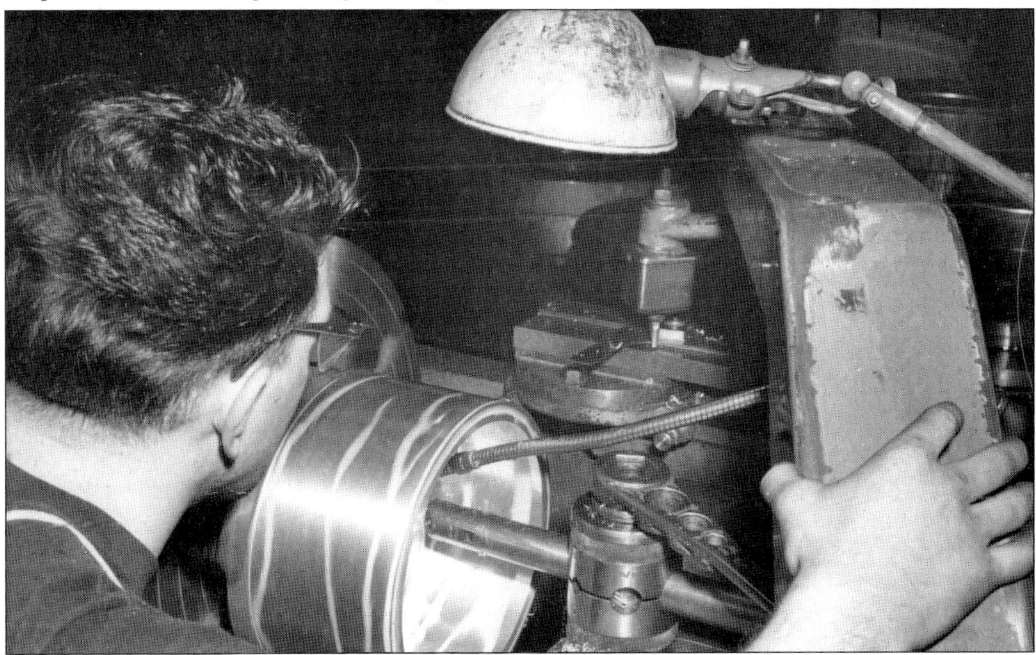

Michael Arndt, part of the West Bend Company Tool and Die Department, bores a sanding chuck for stainless steel bowls in 1967. The Tool and Die Department was as much involved in the preproduction of products as it was in production. Before a product was approved, a test run was made. Many times the dies from the test were tweaked before final approval.

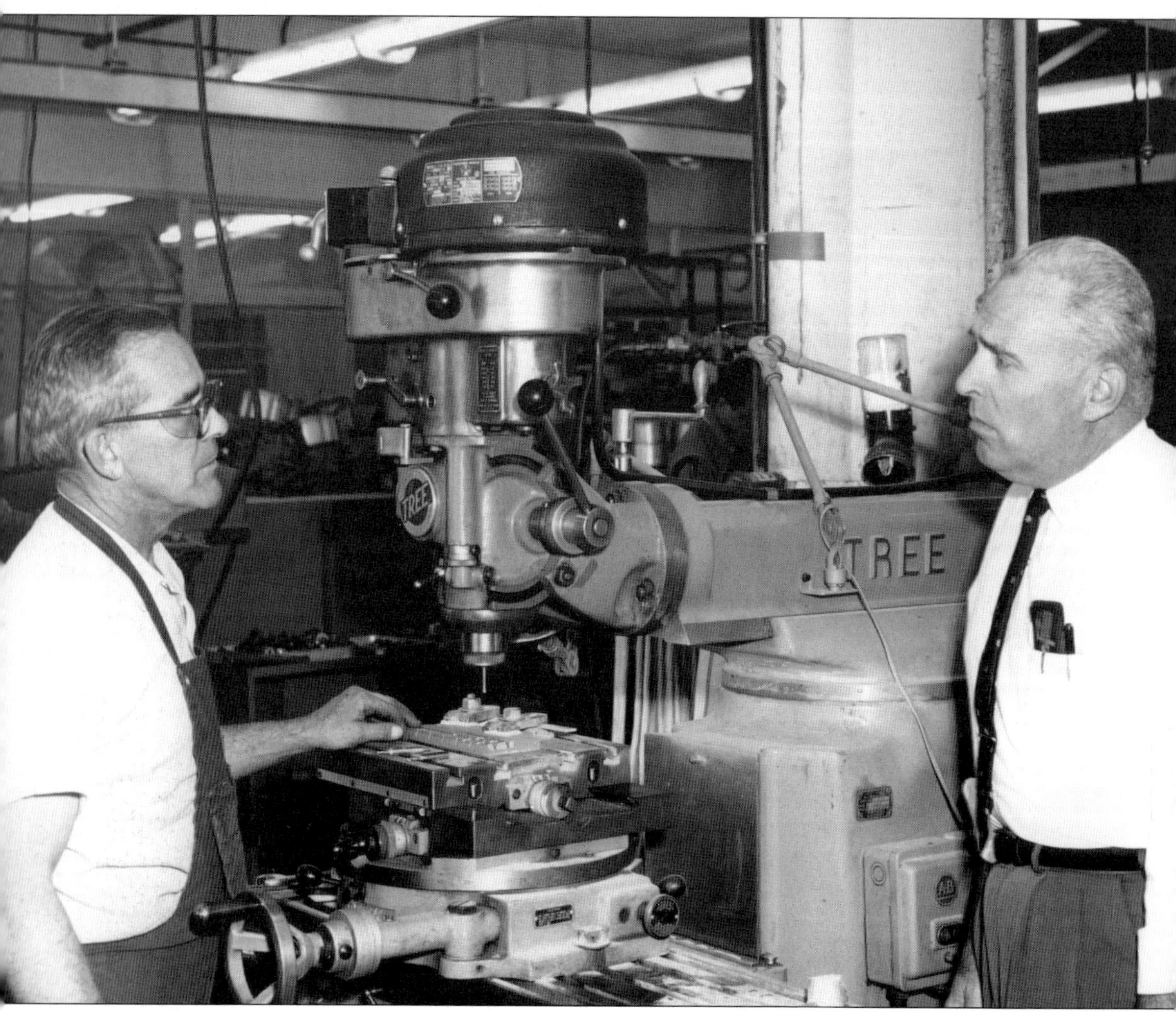

Model maker George Rabb (left) speaks to Bill Laabs, supervisor of Development Engineering for the West Bend Company, in 1968. Both men worked in the Development Engineering Department. The model shop had six full-time model makers, who created samples of new products. The samples were evaluated on design and functionality before being scrapped or put into production.

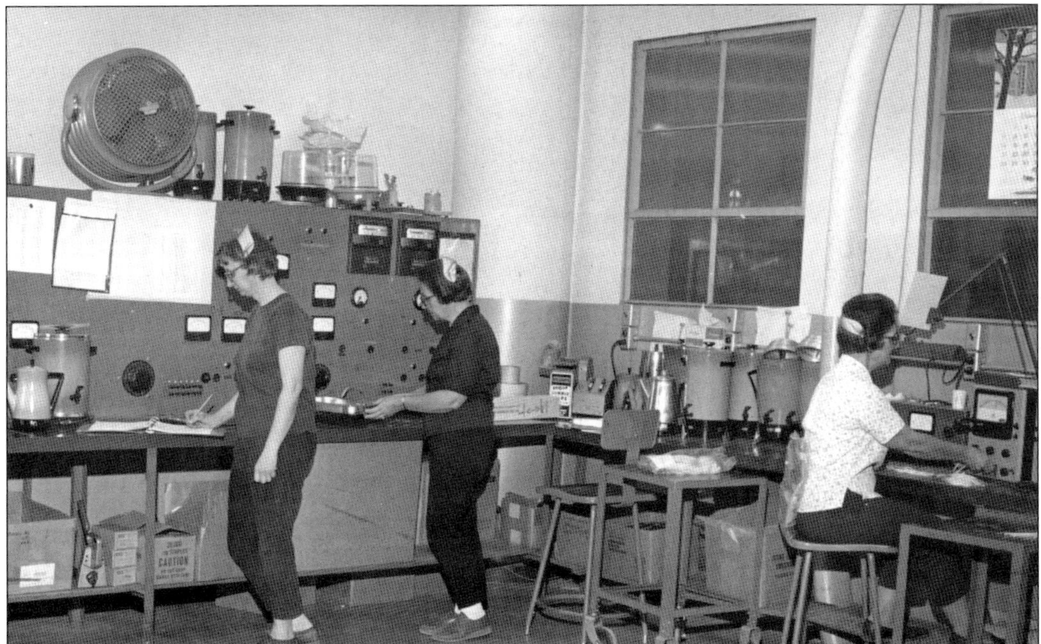

The West Bend Company maintained high quality goods though a very thorough Inspection Department. From left to right, Grace Gieben, Loretta Bastian, and Laverne Miller are testing coffee percolator parts for defects. The department took periodic samples from the line around the clock. If flaws were discovered, production stopped until the cause could be corrected.

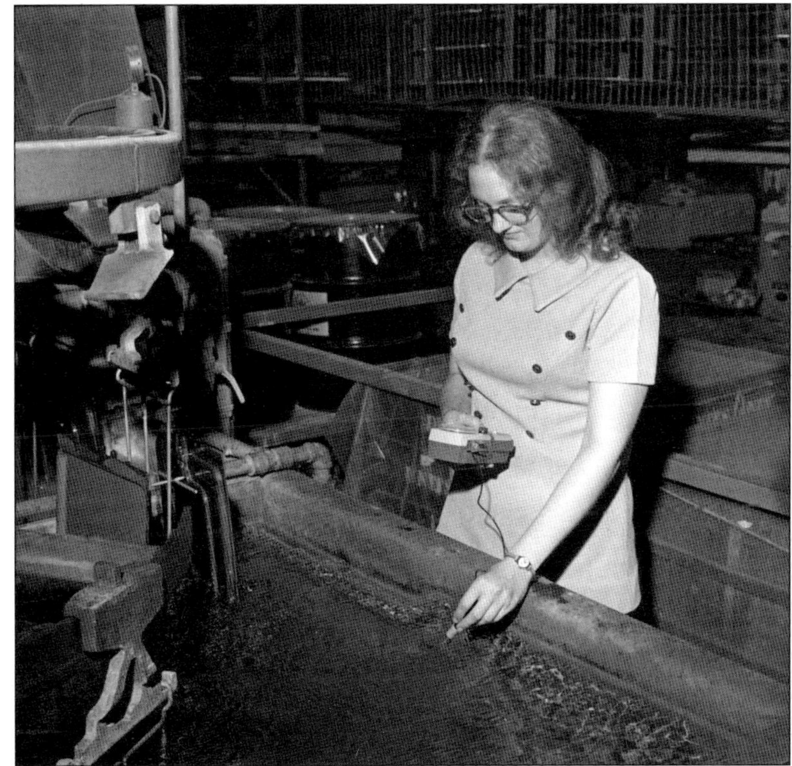

Women worked in all capacities at the West Bend Company. Here Barbara Mommaerts, chemical technician, tests a liquid. In 1972, women were also in other positions typically seen as male jobs. Heavy press operators, security, analyst programmer, attorney, and supervisory positions were all staffed by women.

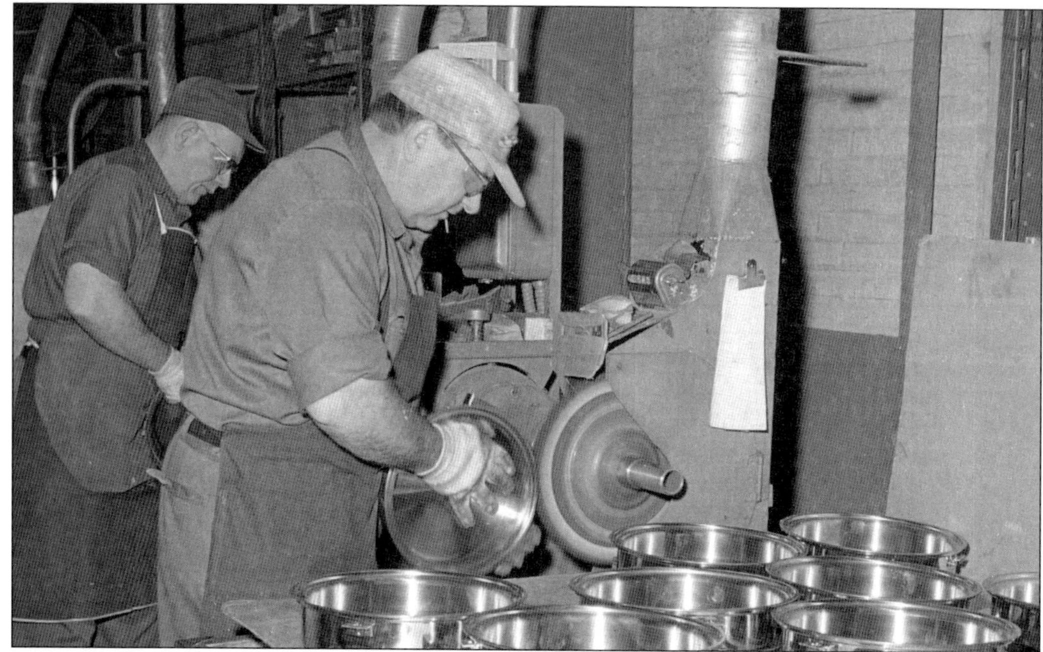

As part of the West Bend Company's Polishing Department, employees Ed Zuelsdorf and Edgar Breuer polish oil-core skillets in 1971. Hand polishing, as seen in this photograph, is done on machines with cloth wheels to create a highly reflective surface. Brushed aluminum, a slightly duller finish, is established by a wire brush wheel.

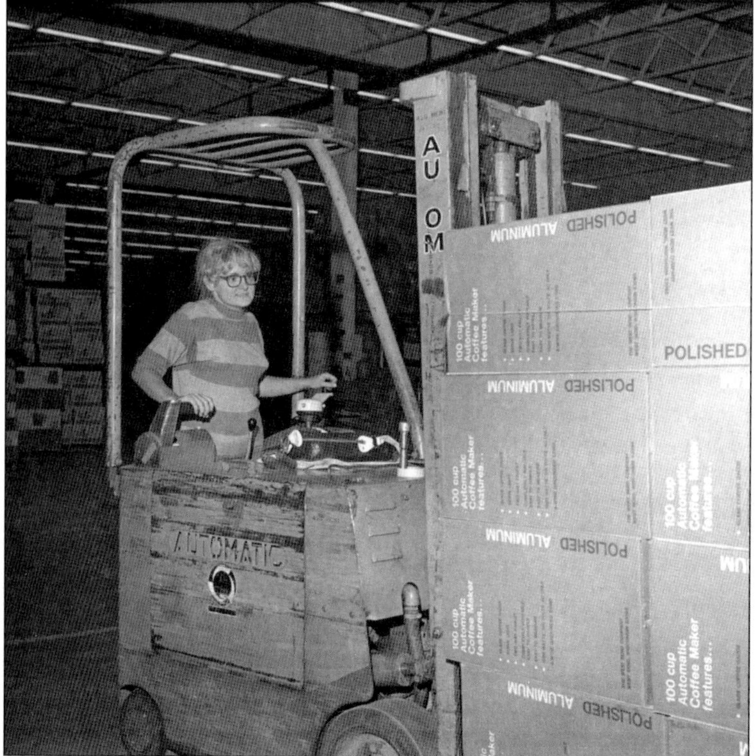

The West Bend Company was proud of its history of being an equal opportunity employer. Since World War II, when women were hired to work on government contracts, women served in many capacities in the West Bend Company that were traditionally seen as male dominated positions. Gayle Poos, seen here, handles stock with a forklift.

This photograph shows Ron Zettler of the West Bend Company's Receiving Department opening a shipment for inspection in 1970. The department worked closely with the Production Control, Purchasing, and Data Processing Departments to ensure shipments were correct and distributed properly.

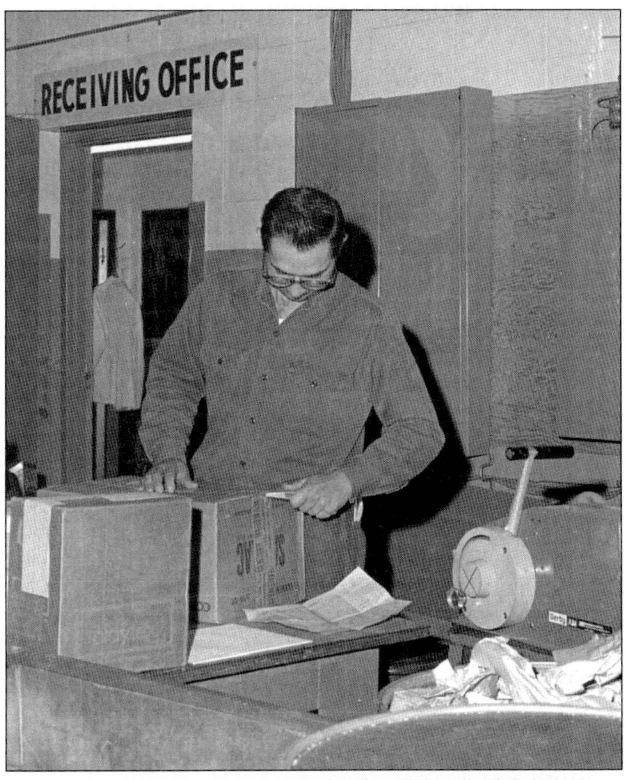

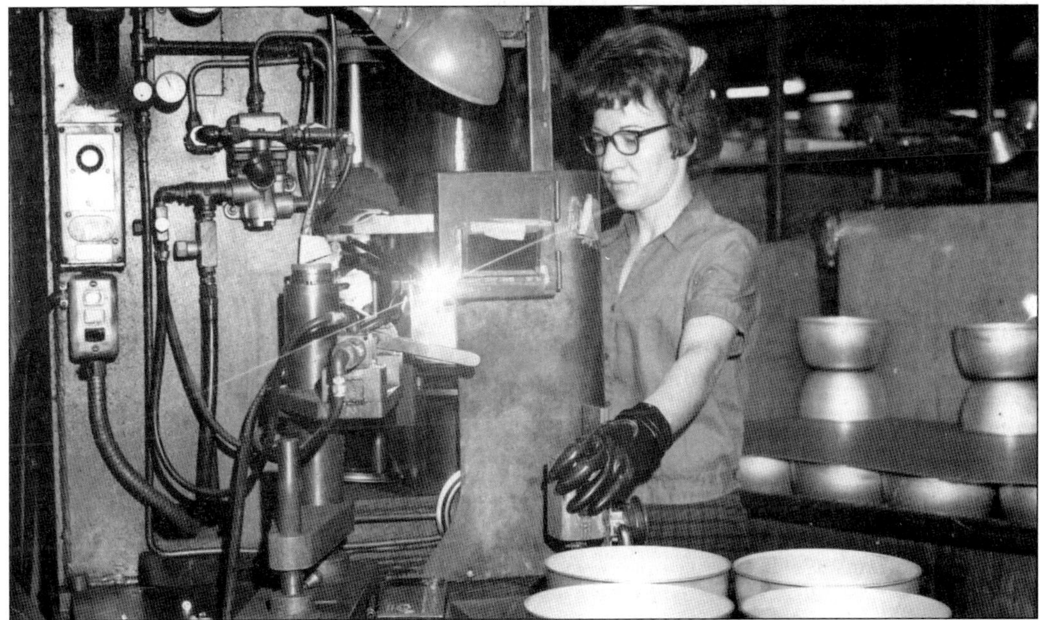

Industrial sales were a large part of the West Bend Company production numbers. Industrial sales produced items for other companies and mixing bowls were one of the largest accounts. The heliarc welding process was created for mixing bowls that attached to an electric mixer. West Bend was the only company in the United States to use the process. The heliarc welding process can be seen in this photograph from 1972.

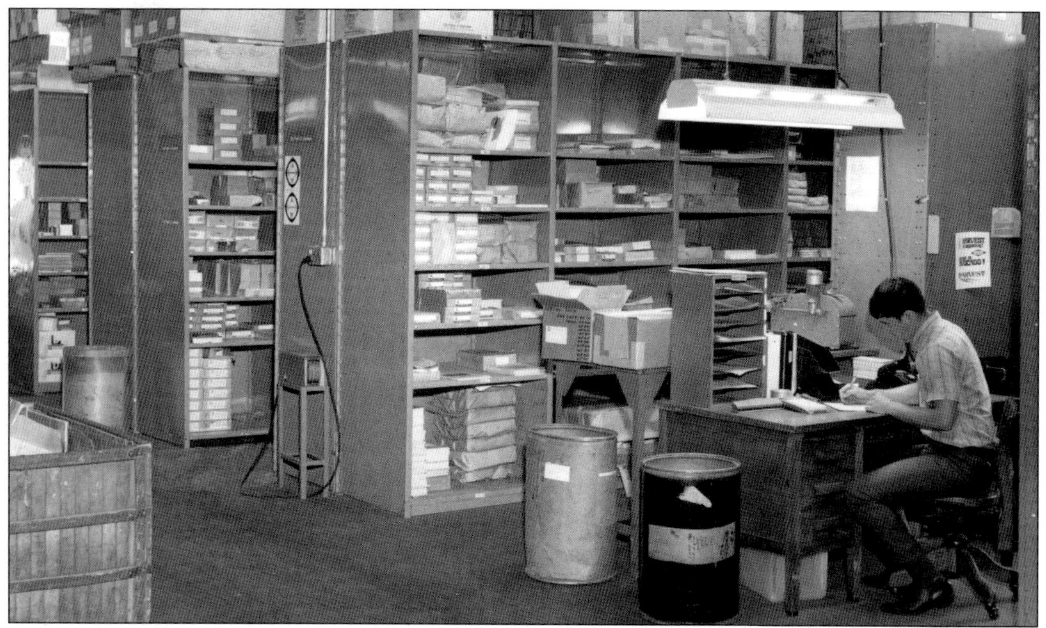

The Customer Service Department of the West Bend Company was there to ensure each customer was a satisfied customer. The department was divided into several sections. The returned goods, parts, and repair sections were in charge of repairing returned or damaged goods. The repack department then shipped the like-new pieces to customers. In this photograph, Mike Hughs sits at his desk in the sales literature section in 1968.

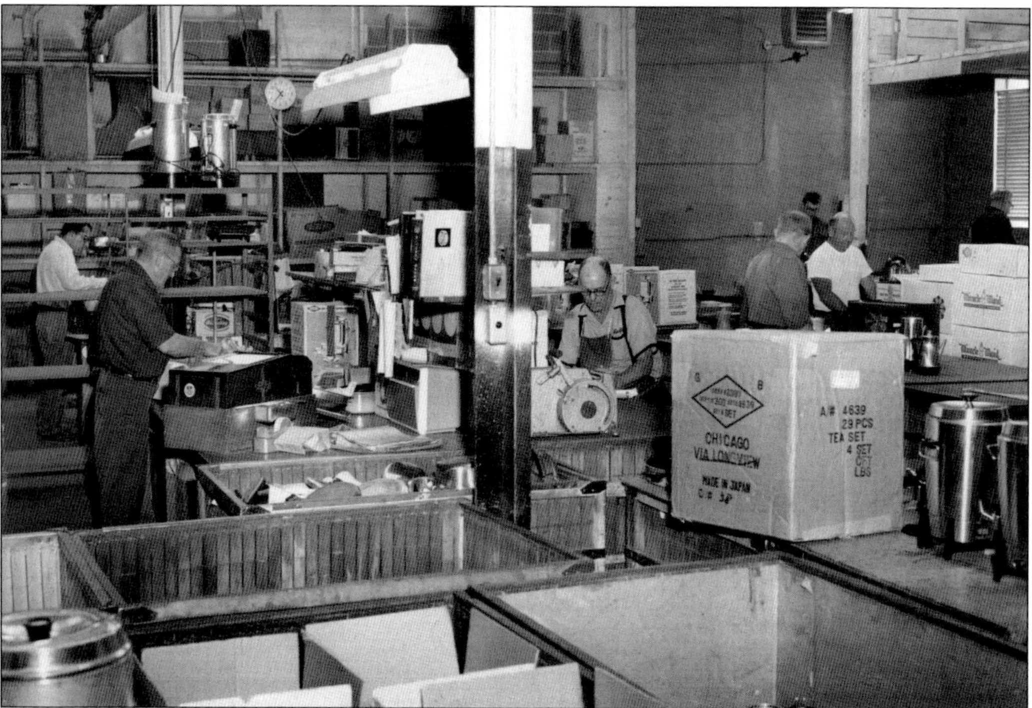

This photograph shows the West Bend Company's returned goods section of the Customer Service Department. Here items returned for repair are evaluated and repaired if possible.

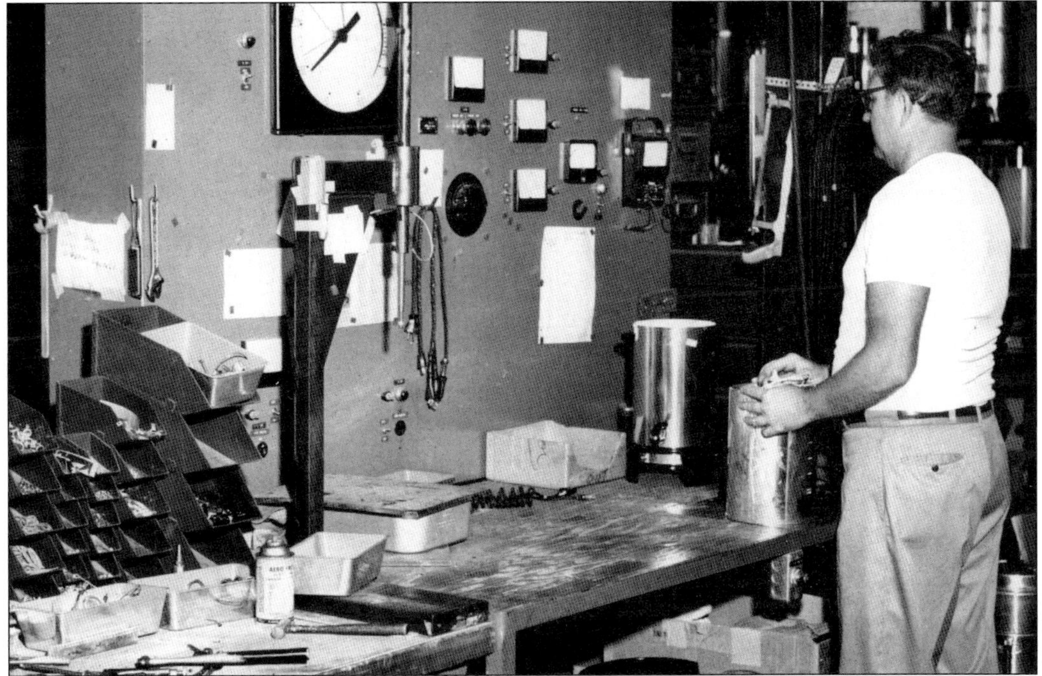

Longtime West Bend Company employee Fred Korth evaluates and repairs a West Bend Company coffee pot. Part of the Customer Service Department, the repair section repaired defective or broken goods.

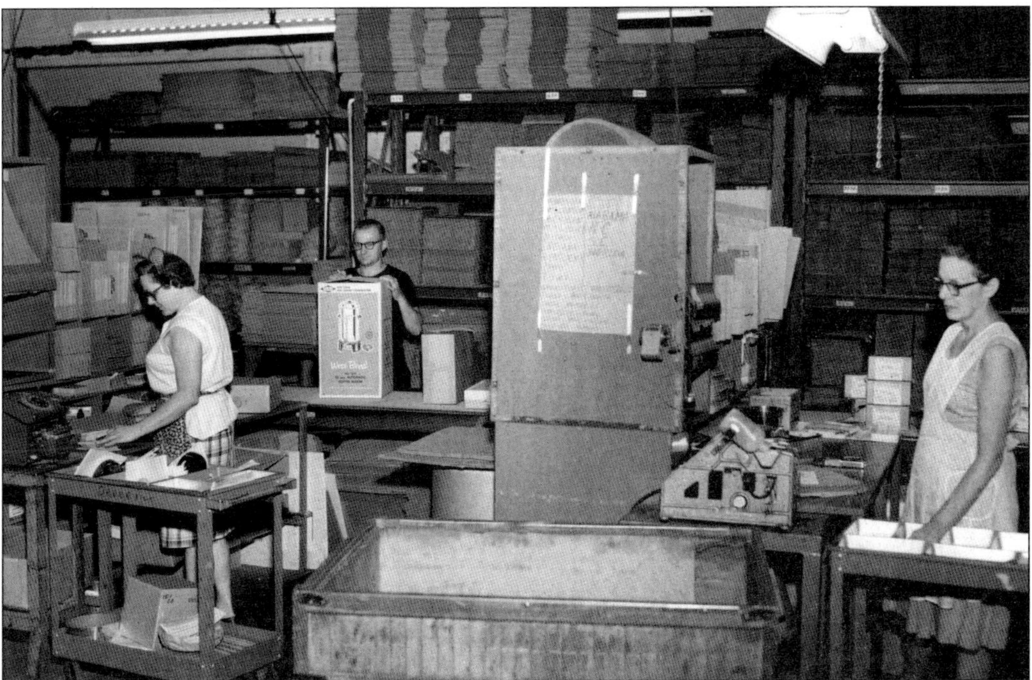

This photograph shows the repack section. When a returned West Bend Company product has been repaired, this division of the Customer Service Department gets it ready for shipment to the customer.

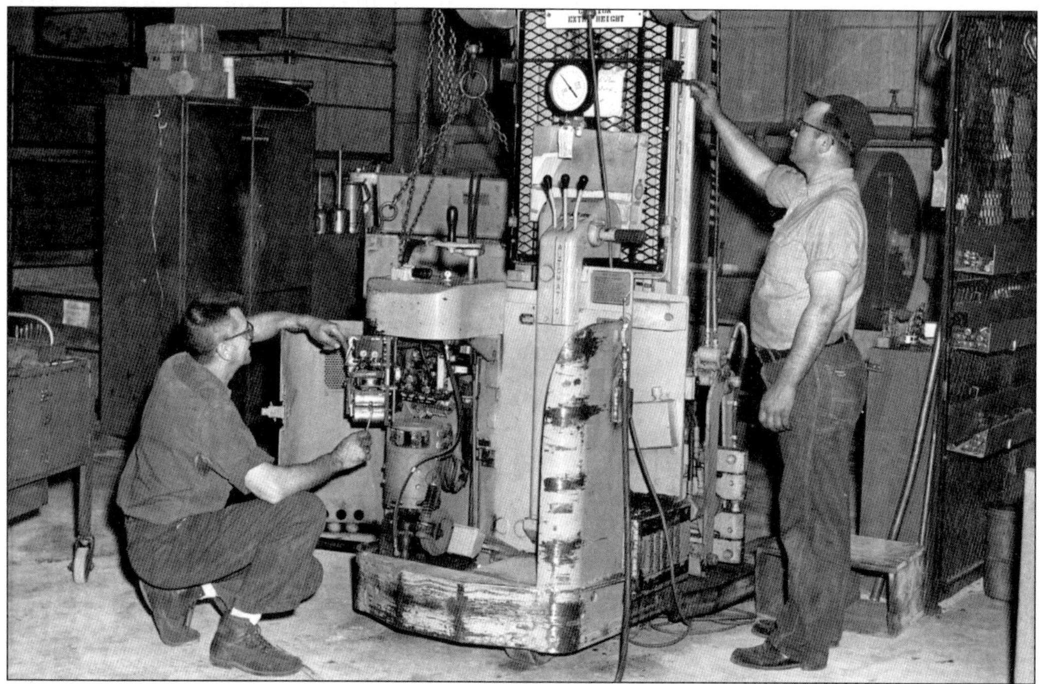

In this photograph from 1967, Elmer Reiser and Buddy Berth repair a forklift. The West Bend Company employed 51 individuals in the machine shop's maintenance division. These men were called upon to do everything from repairing manufacturing equipment to hanging awnings.

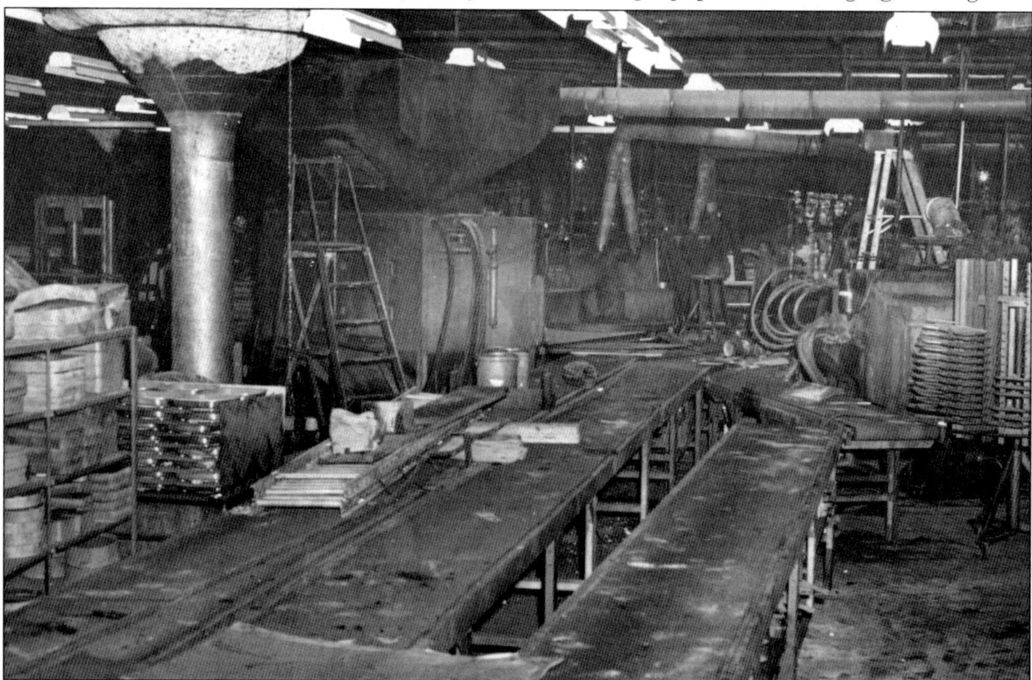

On Friday, May 13, 1966, at around 8:30 p.m., a fire broke out in the Teflon Coating Department of the West Bend Company. Trained company firemen aided the fire department in containing the fire. This photograph shows the department shortly after the fire was put out.

Joe Marx removes damaged concrete in the Teflon Coating Department after the fire. Although the fire spread quickly, fire doors contained the damage to other areas. There were no injuries, and fast repair work restored operations in less than a week. This type of work is one example of the wide variety of tasks taken on by the Maintenance Department.

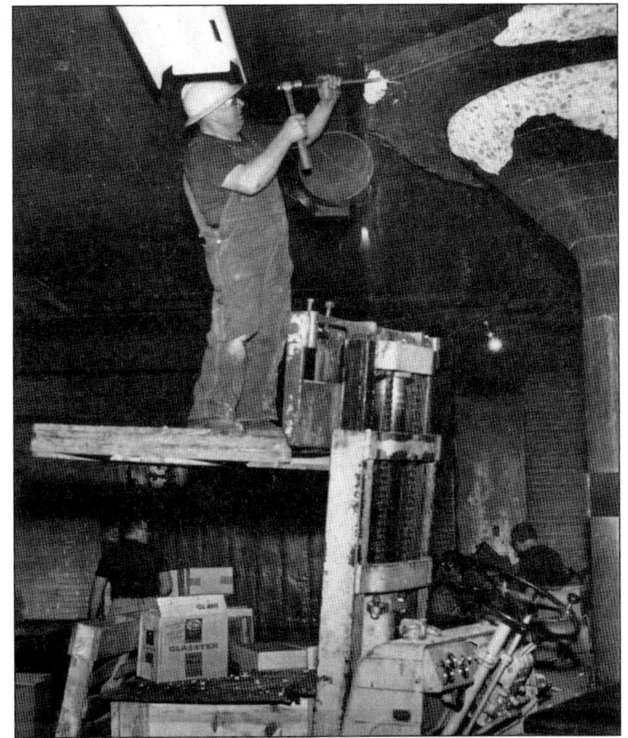

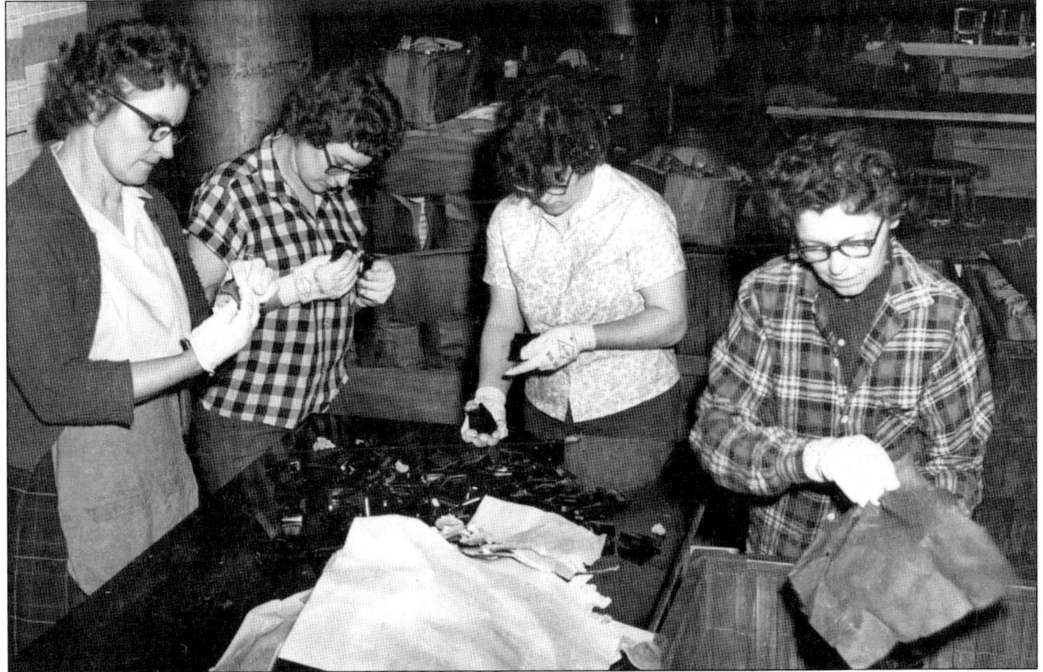

Although the fire in the Teflon Coating Department at the West Bend Company was contained, it did cause extensive damage to equipment and products. From left to right, Virginia Schloemer, Phyllis Hagedorn, Layanne Hesprich, and Helen Cook are seen sorting though handles damaged by the fire.

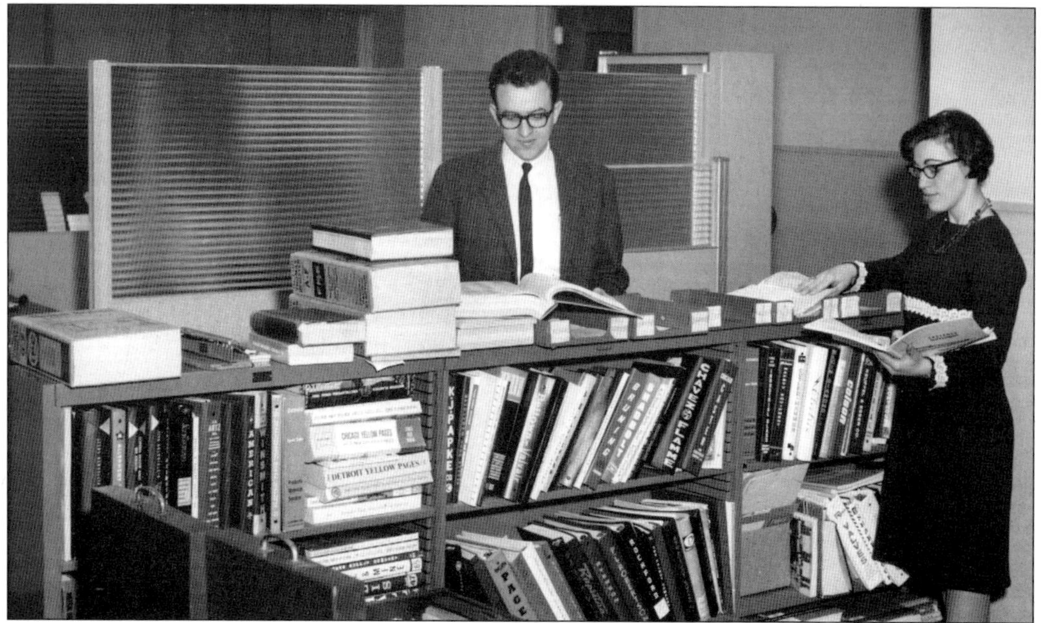

A typical scene in the West Bend Company's Purchasing Department library in 1967, buyer Ron Teofilo and secretary Sharon Laufer use the library to find materials and supplies. Purchasing was in charge of buying everything that was needed throughout the company. To facilitate this immense job, the department was divided into five sections, each specializing in specific products.

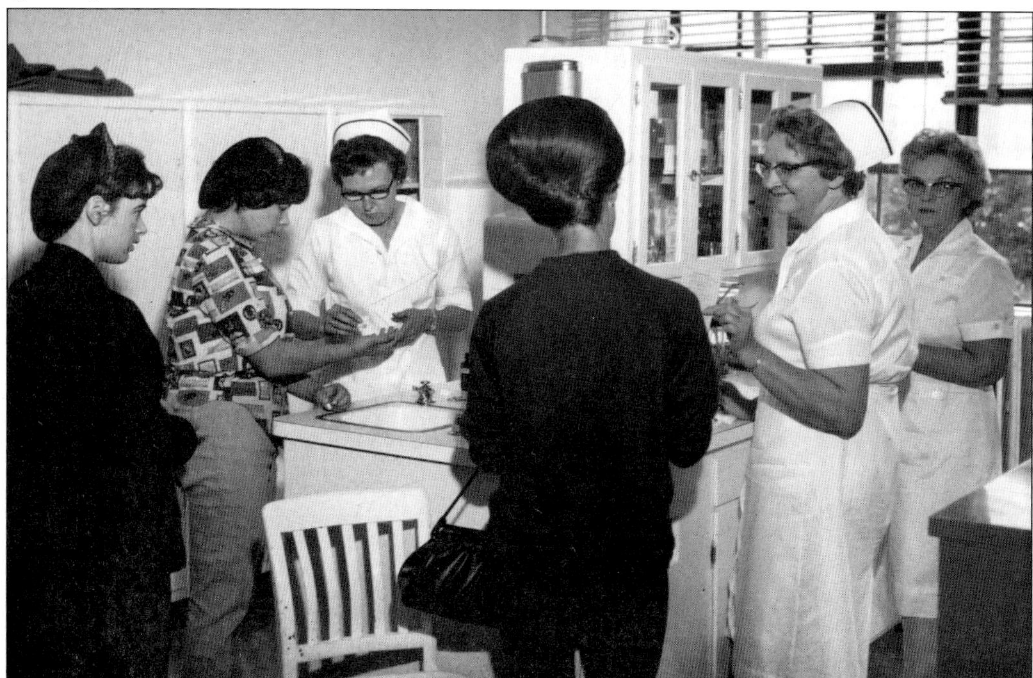

This photograph shows the first aid room in action. A part of the Personnel Department, the nurses on staff attended to first aid needs and screened new employees. From left to right are nurses Ruth Fairly, Mildred Hoffman, and Elsie Thorson.

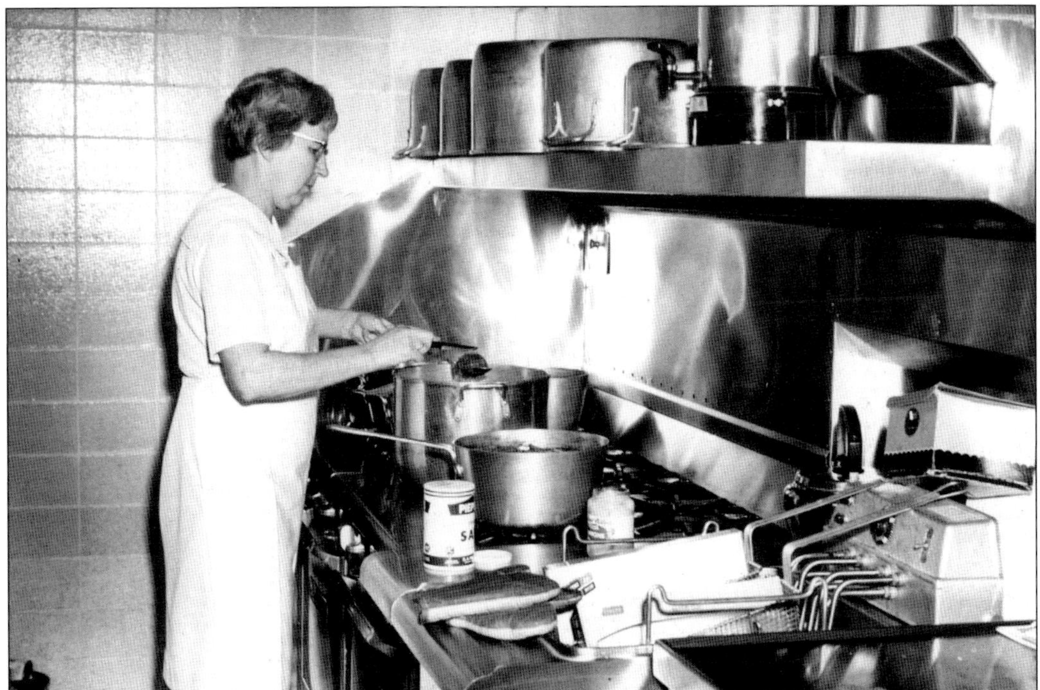

Mary Bohn, an eight-year cafeteria veteran, prepares the noon lunch for workers in 1967. The West Bend Company cafeteria prepared meals for 90 to 140 people every work day.

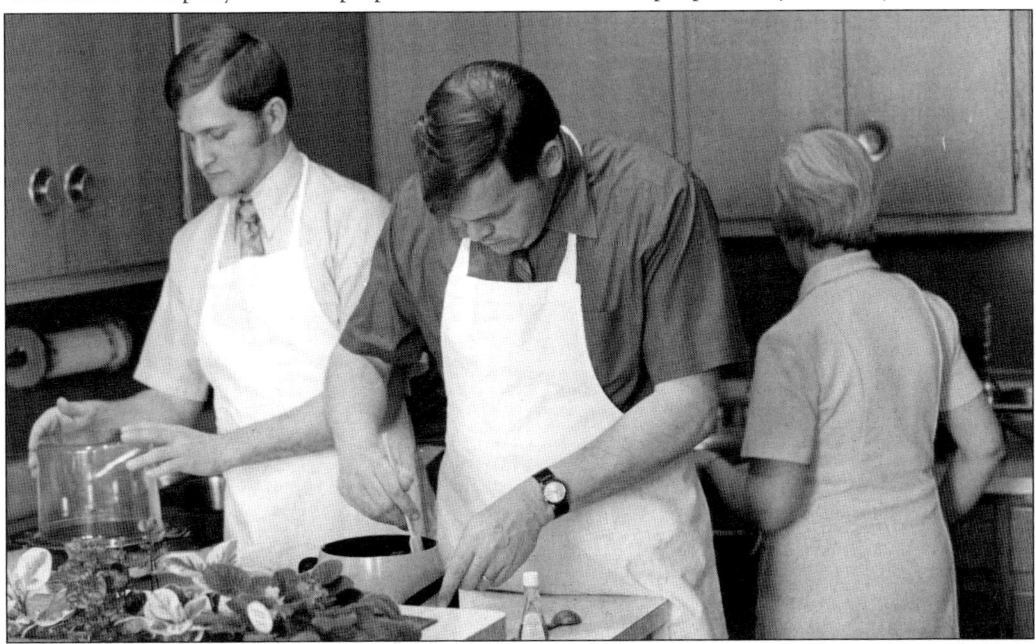

The West Bend Company test kitchen served as a place to train sales personnel. New and old recruits alike spent time under the guidance of Jeanette Weinberger in 1971 to learn about the products they were peddling. Using the product allowed the salesmen and women to sell with firsthand experience. Some sales pitches were also used in home visits, or direct sales. These required the salesperson to cook for the client and be proficient with the product.

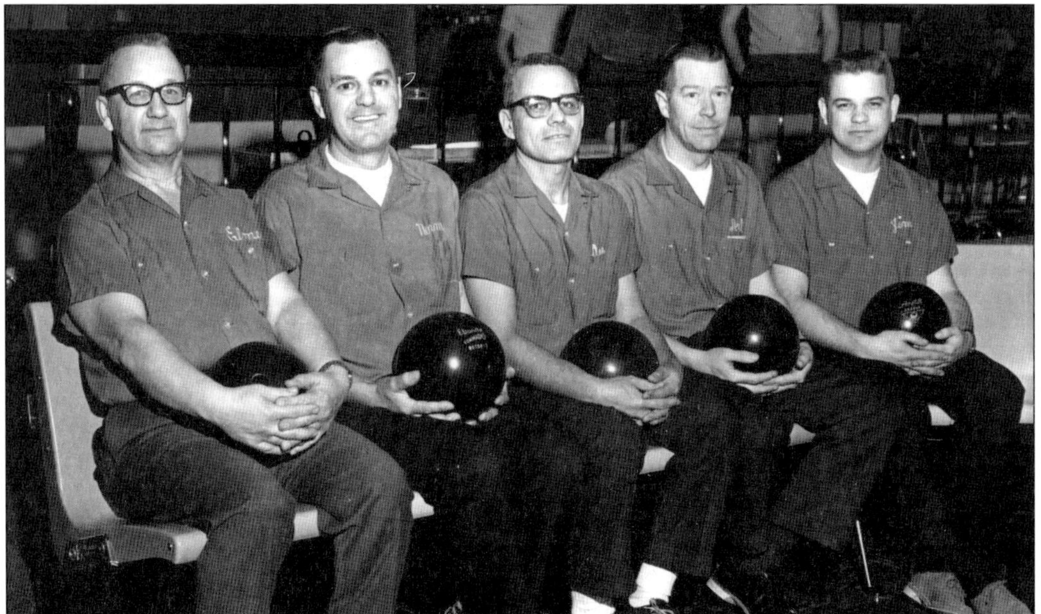

The West Bend Company supported employee sports teams. Here some of the members of the bowling team entered in the local bowling alley Weiland's All-Star League in 1966. From right to left they are Elmer Harter of personnel, Norm Minsky of development engineering, Dee Goltry of product engineering, Art Schmitt of development engineering, and Jim Wondergem of product engineering.

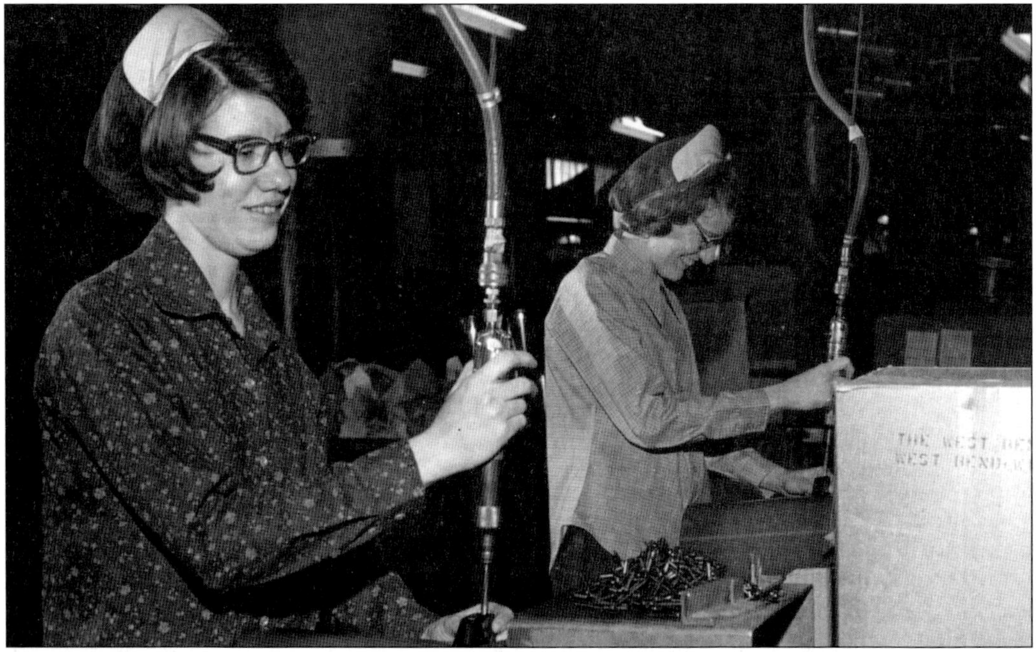

Caroline (left) and Anastasia Kiefer operate automatic screwdrivers in 1970. The Kiefer family is one of the examples of the pride West Bend Company employees took in their work. The company promoted family involvement, and in 1970, six members of the Kiefer family were working in the West Bend plant.

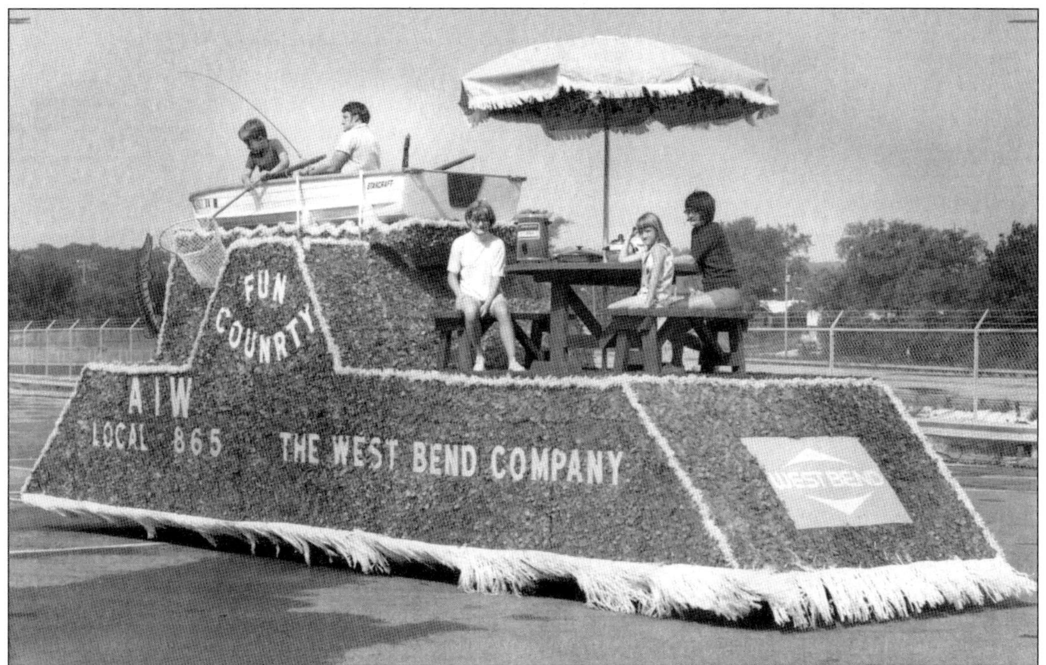

The Allied Industrial Workers Union Local 865 was an annual feature in the West Bend Labor Day parade. The float was a fun project for the workers and families and also served to advertise West Bend Company products. Their 1971 parade float featured several kitchen utensils.

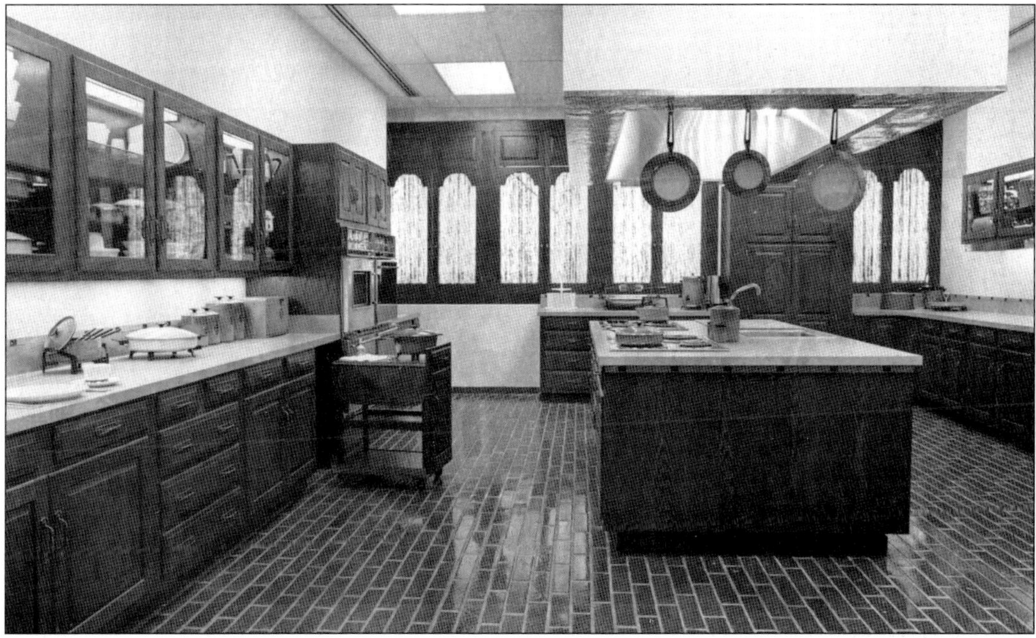

In 1968, West Bend Company merged with Rexall Drug and Chemical Company, which later became Dart Industries. This photograph from 1970 shows the Dart Industry's Dream Kitchen at the company headquarters in Beverly Hills, California. The kitchen was able to serve 45 people while displaying and using the products. Like its predecessors at the West Bend headquarters in Wisconsin, this kitchen served to promote the goods through example.

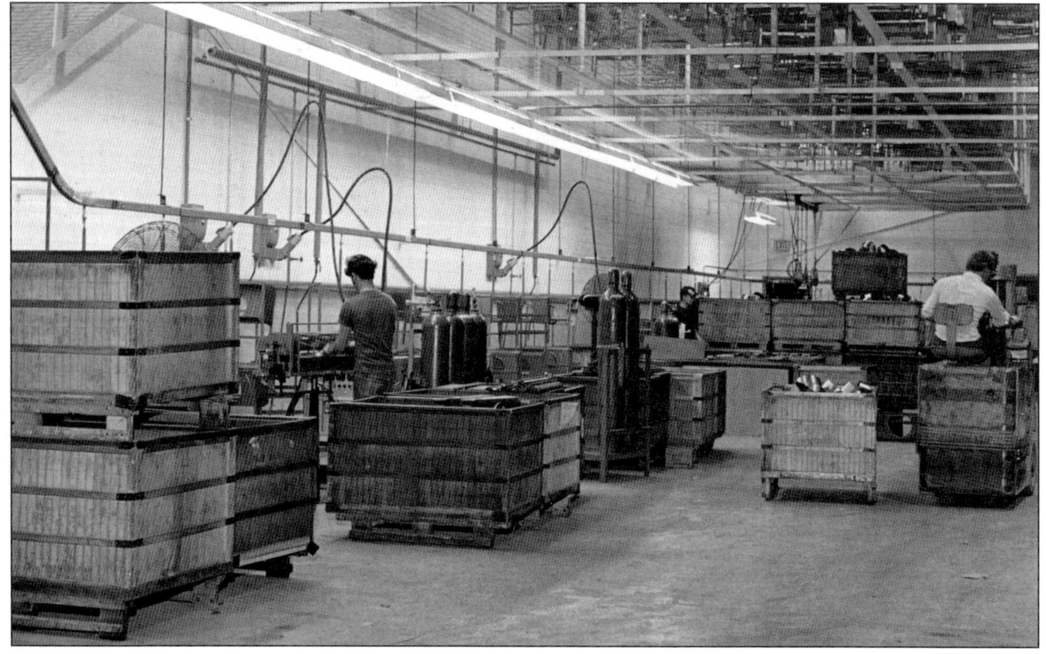

The West Bend Company East Plant, located on Schmidt Road in West Bend was built in 1968. It was built as a modern distribution center and also housed additional manufacturing and office space. The section of the plant shown here was dedicated to the welding of humidifier chassis and components.

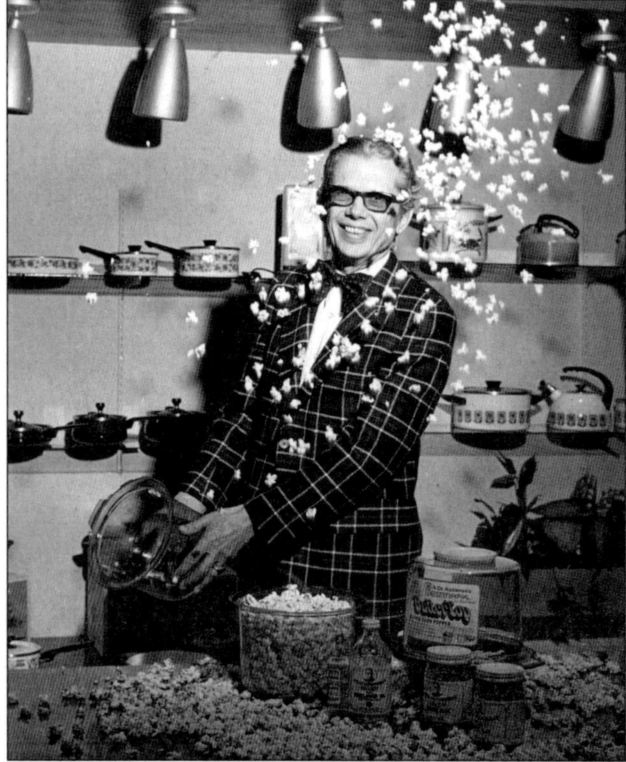

Orville Redenbacher tosses his popcorn in a publicity photograph for the West Bend Company around 1970. West Bend Company often partnered with products, such as popcorn, to advertise their merchandise.

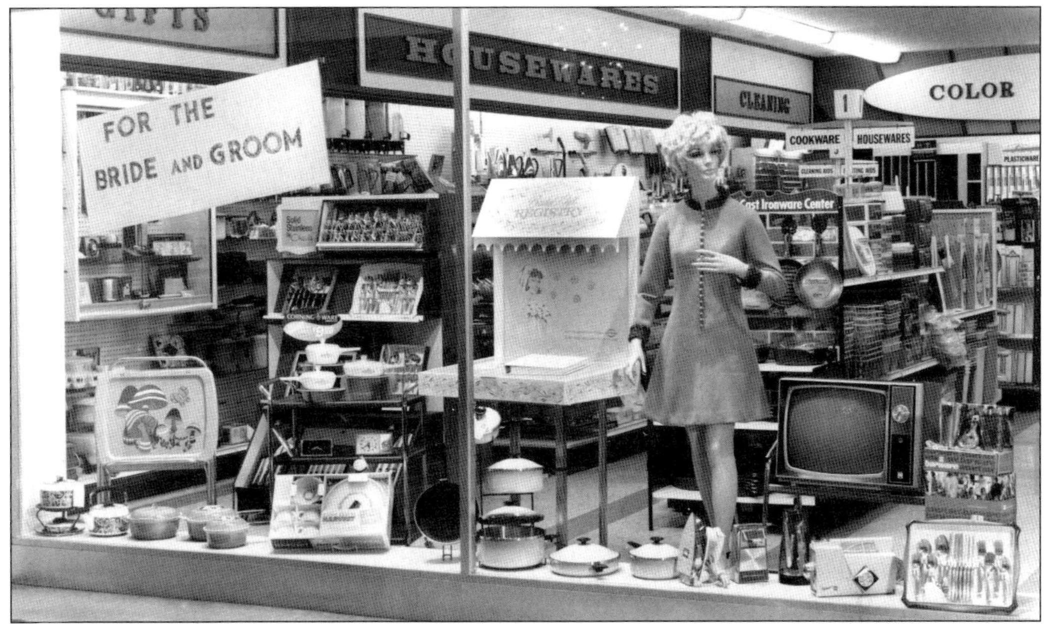

A bridal registry window display from around 1970 shows one of the ways companies advertised to niche markets. One of the strongest areas of marketing the West Bend Company employed was for bridal gifts. Displays such as this one were usually set up by regional West Bend salesmen from standard kits sent out by the company. Registers were kept in a binder alphabetized by the bride's last name.

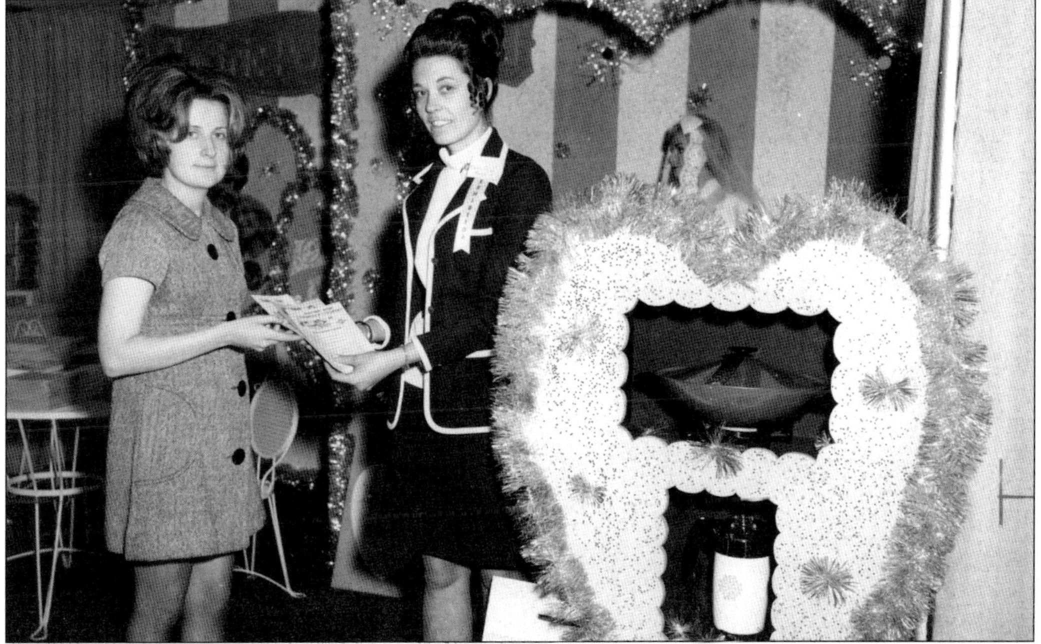

The West Bend Company's success at becoming a household name was helped along at bridal shows like this one in 1972. Here Marie Smith, the home economist for the Chattanooga Electric Power Board, passes out recipe booklets for the West Bend electric wok at a display advertising electric appliances in Chattanooga, Tennessee.

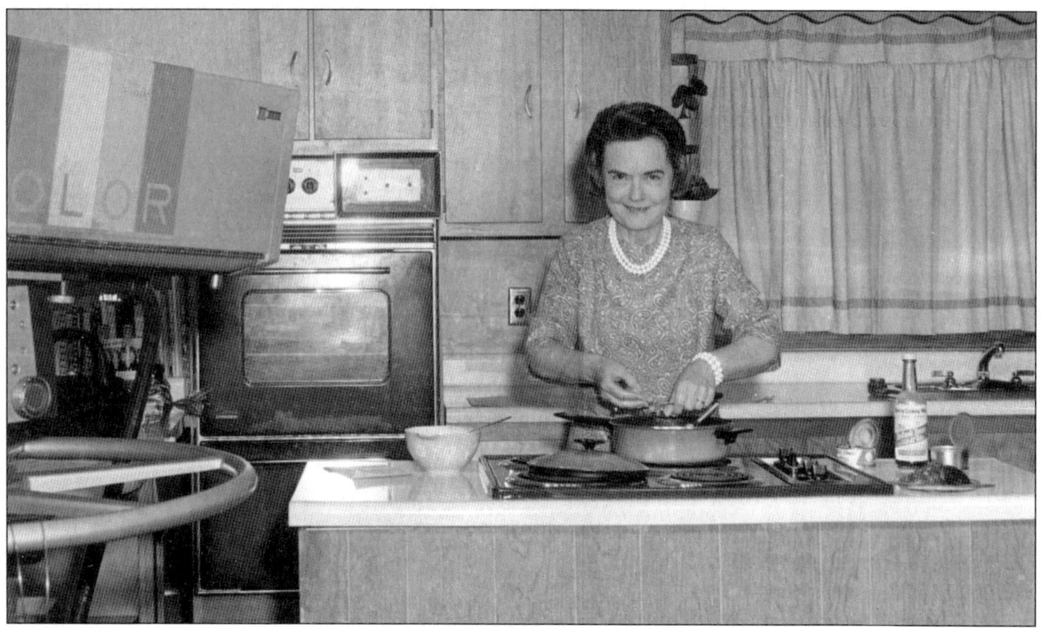

In 1967, Mary Starr, home economist for television station WATE-TV in Knoxville, Tennessee, used the avocado green Country Inn cookware in a demonstration on her show. She wrote to the company to say how much she enjoyed using the cookware and how well the avocado showed up on color broadcasts. It was national exposure such as this that helped to popularize the color and brand.

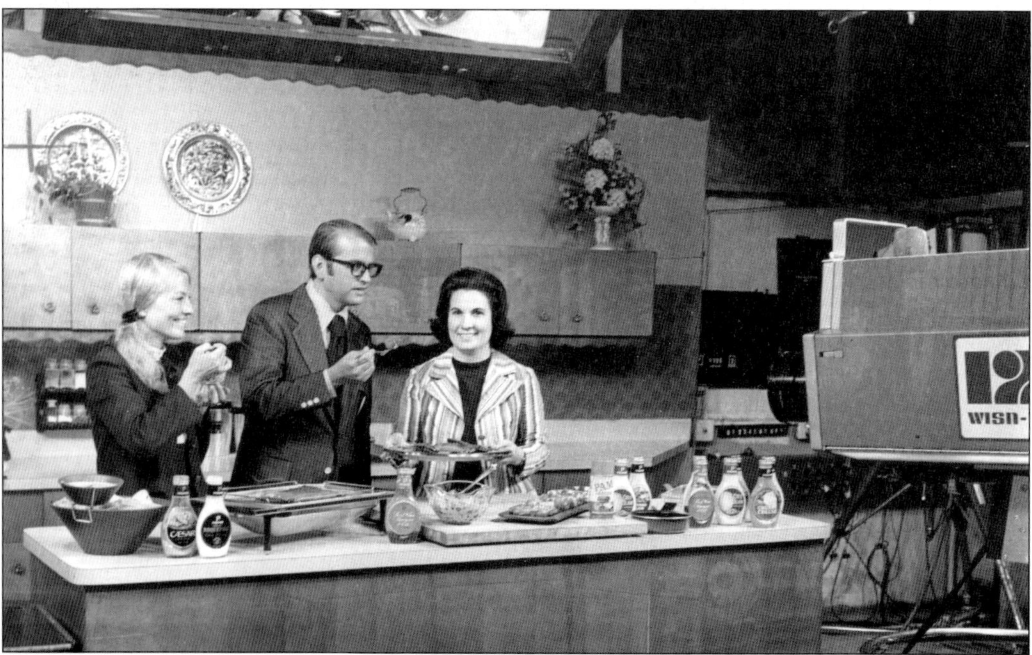

One of the most important advertising outlets for the West Bend Company was television. Here Bunny Rasasch and Bruce Bennett, cohosts of *Dialing for Dollars* on WISN-TV in Milwaukee, taste a meal prepared on the West Bend broiler-rotisserie on July 24, 1972. The home economist doing the serving is Sophie Kay.

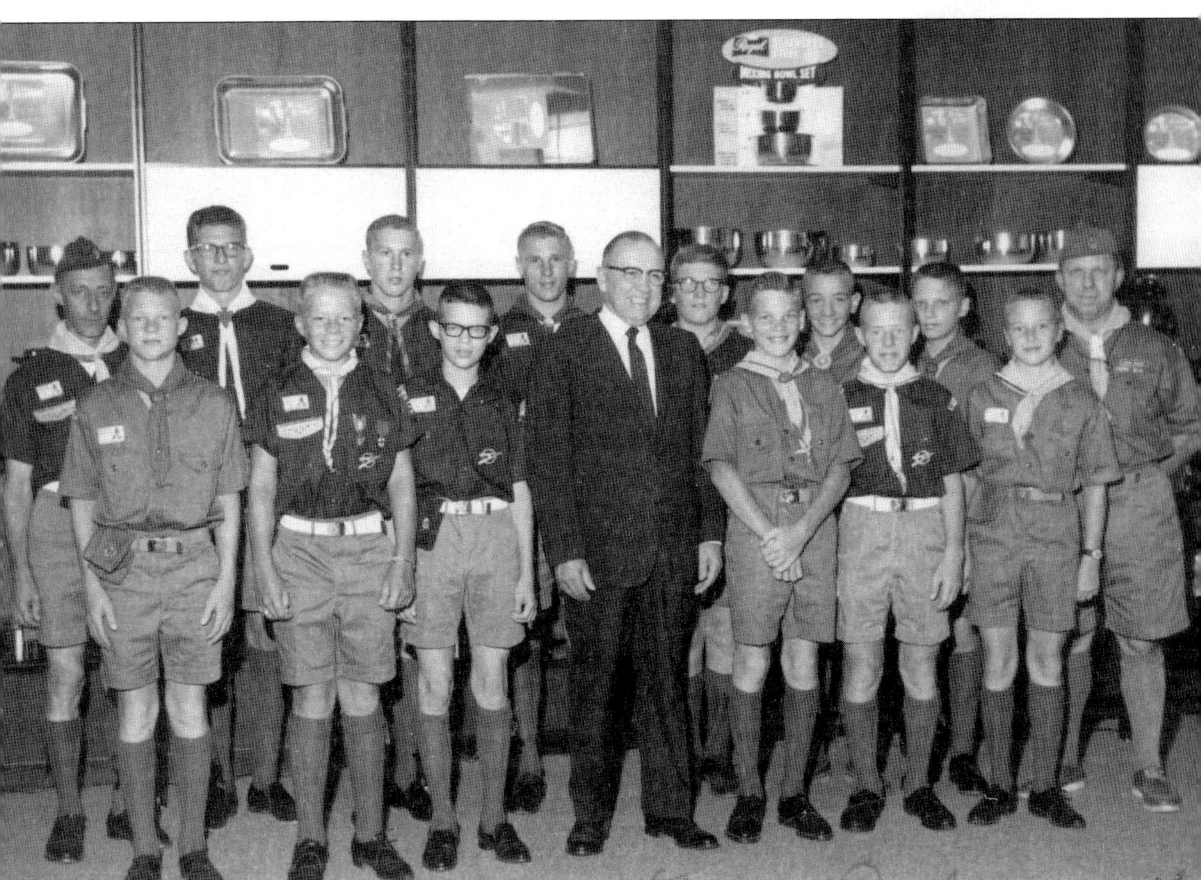

Regal Ware Inc. began producing cook kits and canteens for the Boy Scouts of America in 1948. The first items were made of tin; they then moved into aluminum and eventually plastic items. Pieces included cookware sets, utensils, and canteens. This photograph, taken in 1964, is of Troop 58 of Kewaskum. The man in the back row on the left is Squinty Manthie. In the center, wearing a suit, is Regal Ware Inc. founder J. O. Reigle. The third boy in from the right in the back row is current president of Regal Ware Inc. and J. O. Reigle's grandson Jeffery Reigle.

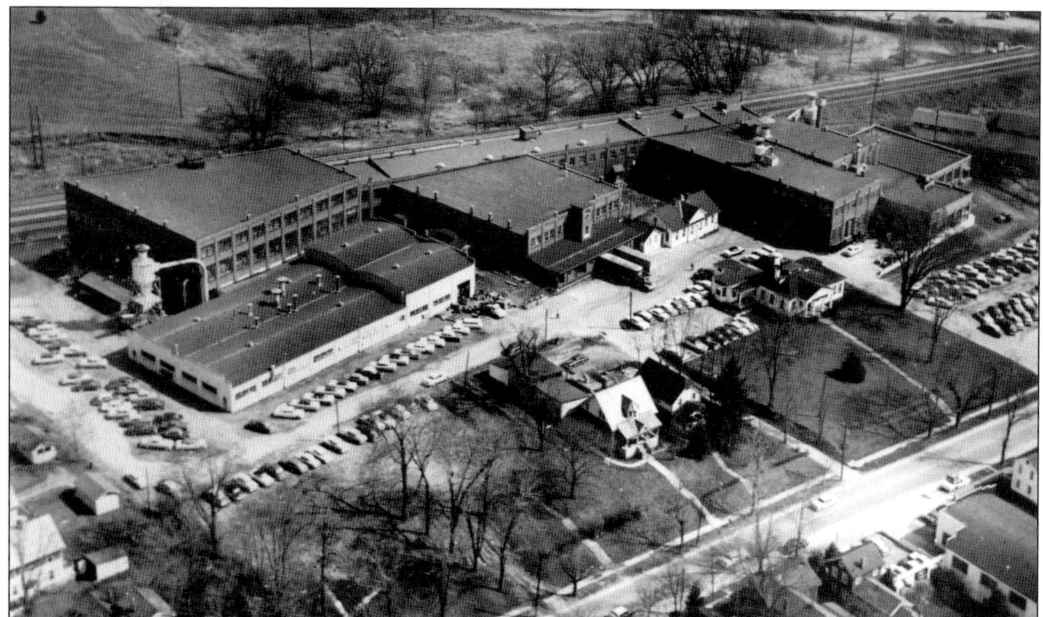

The Buckeye Division of Mardigian Corporation in Wooster, Ohio, was acquired by Regal Ware Inc. in March 1964. The Buckeye Division produced aluminum and stainless steel cookware, small electrical appliances, and other household goods. Regal Ware Inc. created a new company, Buckeye Ware Inc., from the acquisition. The facilities in Ohio were used to aid marketing, manufacturing, warehousing, and distribution of all of Regal Ware Inc.'s products.

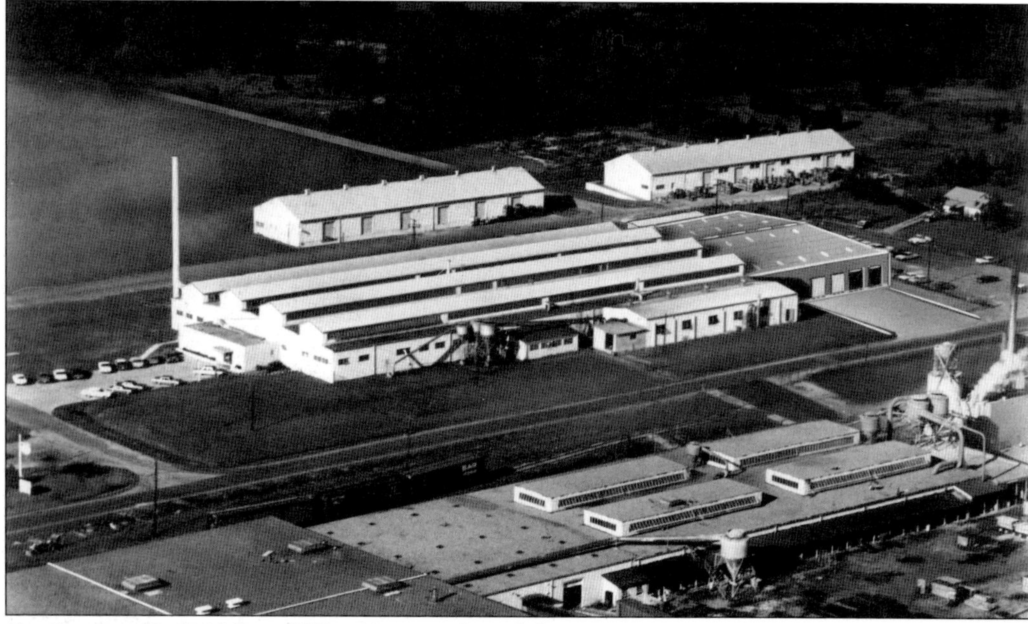

On October 18, 1964, Regal Ware Inc. purchased the Norris Industries plant in Flora, Mississippi. The plant manufactured cookware and was purchased to provide production relief to the Kewaskum plant. The Mississippi plant became known as Regal Southern Inc. and allowed Regal Ware Inc. to enter into copper clad, aluminum clad, and Teflon finish markets. This aerial view of the Mississippi factory was taken in 1964.

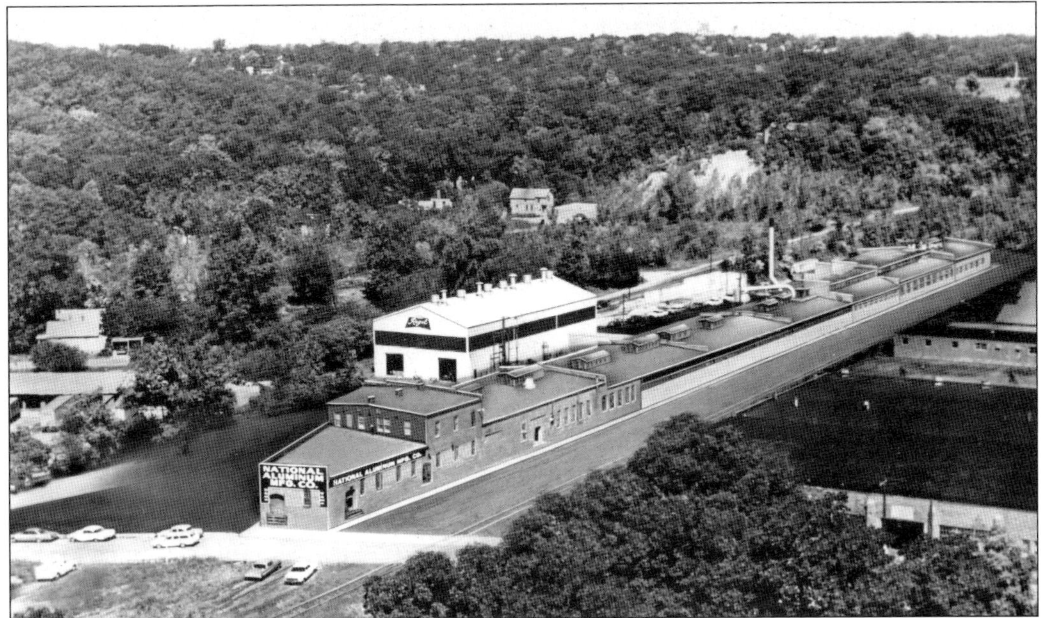

In 1969, Regal Ware Inc. acquired a controlling interest in the National Aluminum Manufacturing Company of Peoria, Illinois. This allowed Regal Ware Inc. to enter into the cast aluminum manufacturing market. National Aluminum Manufacturing Company was in business for 50 years before the transfer occurred. This aerial view of the National Aluminum Manufacturing Company was taken around 1969.

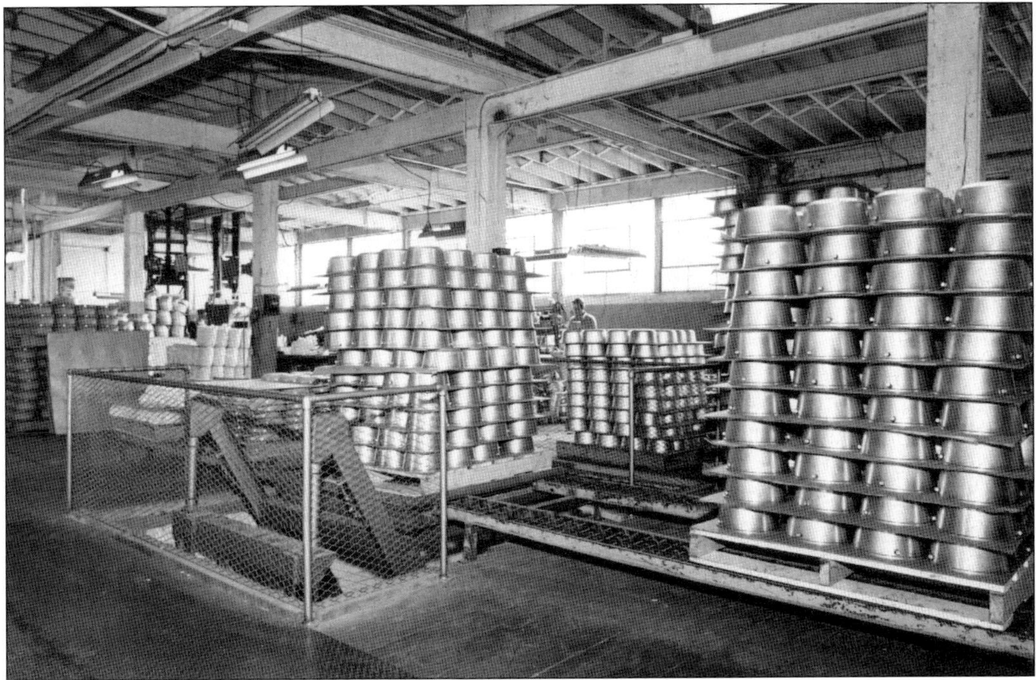

In 1969, Regal Ware Inc. made the purchase of the National Aluminum Manufacturing Company in Peroia, Illinois. The purchase of National allowed Regal Ware Inc. to enter the cast aluminum market. This photograph from 1970 shows product stacked at the Peoria plant.

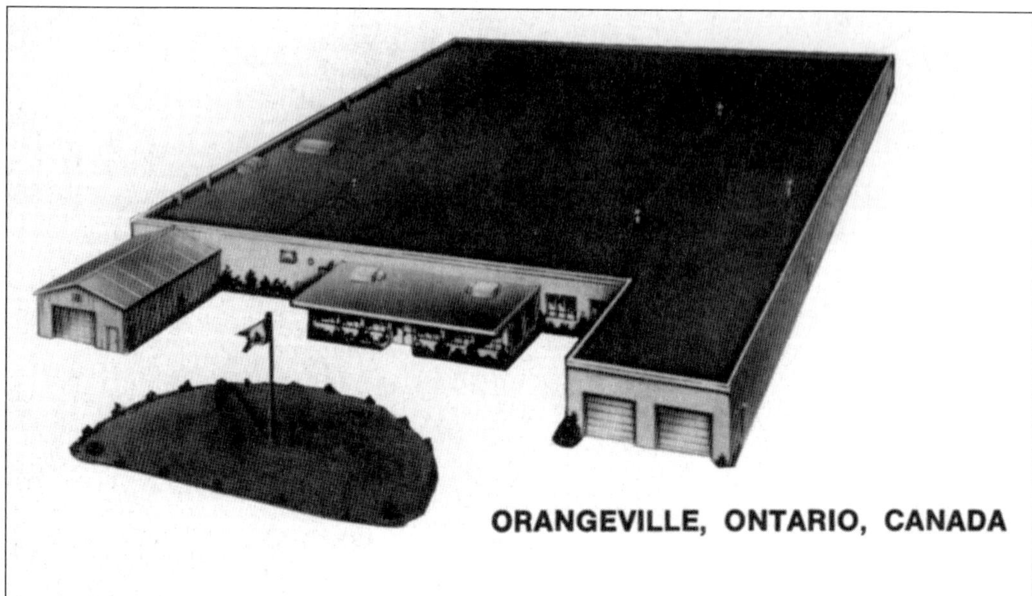

This artist conceptual drawing depicts the Orangeville, Ontario, Canada, plant of Regal Ware Inc. In 1972, the company purchased the plant formally operated by Sunbeam. Regal Ware Inc. manufactured stainless steel products and used the factory as a distribution center. One of the main reasons for choosing the plant was its central location, which sped up distribution.

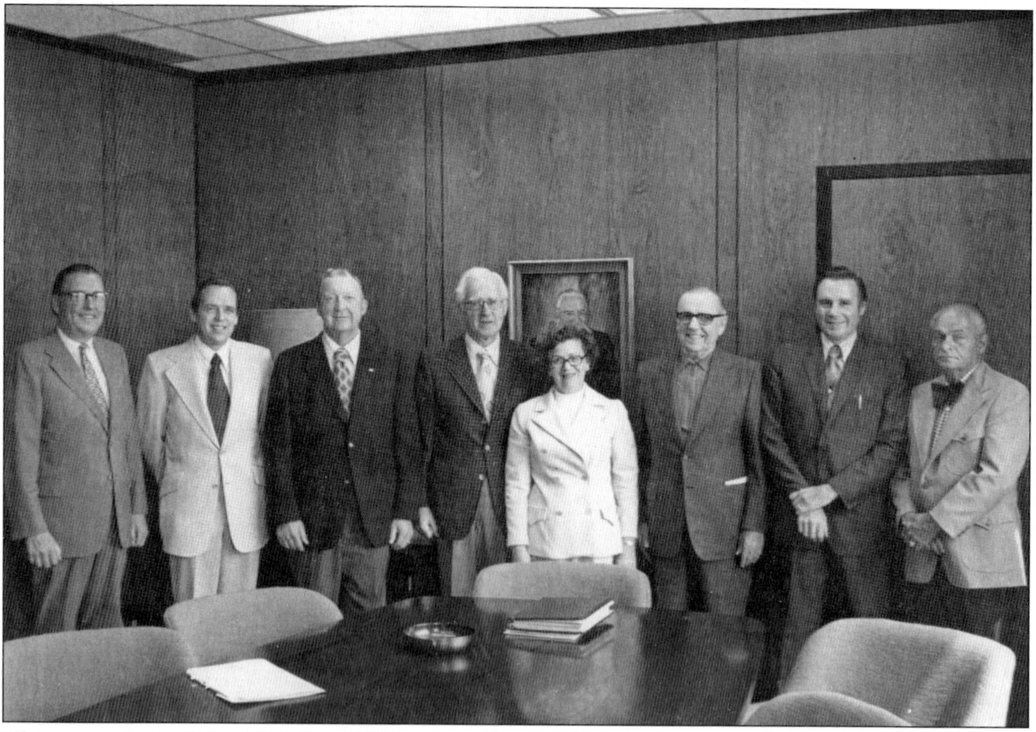

This portrait is of the 1971 Regal Ware Inc. Board of Directors. From left to right are James D. Reigle, Richard Reigle, Milton Meister, L. N. Peterson, Edna Oster, J. O. Reigle, Ronald Reigle, and Ned Sengpiel.

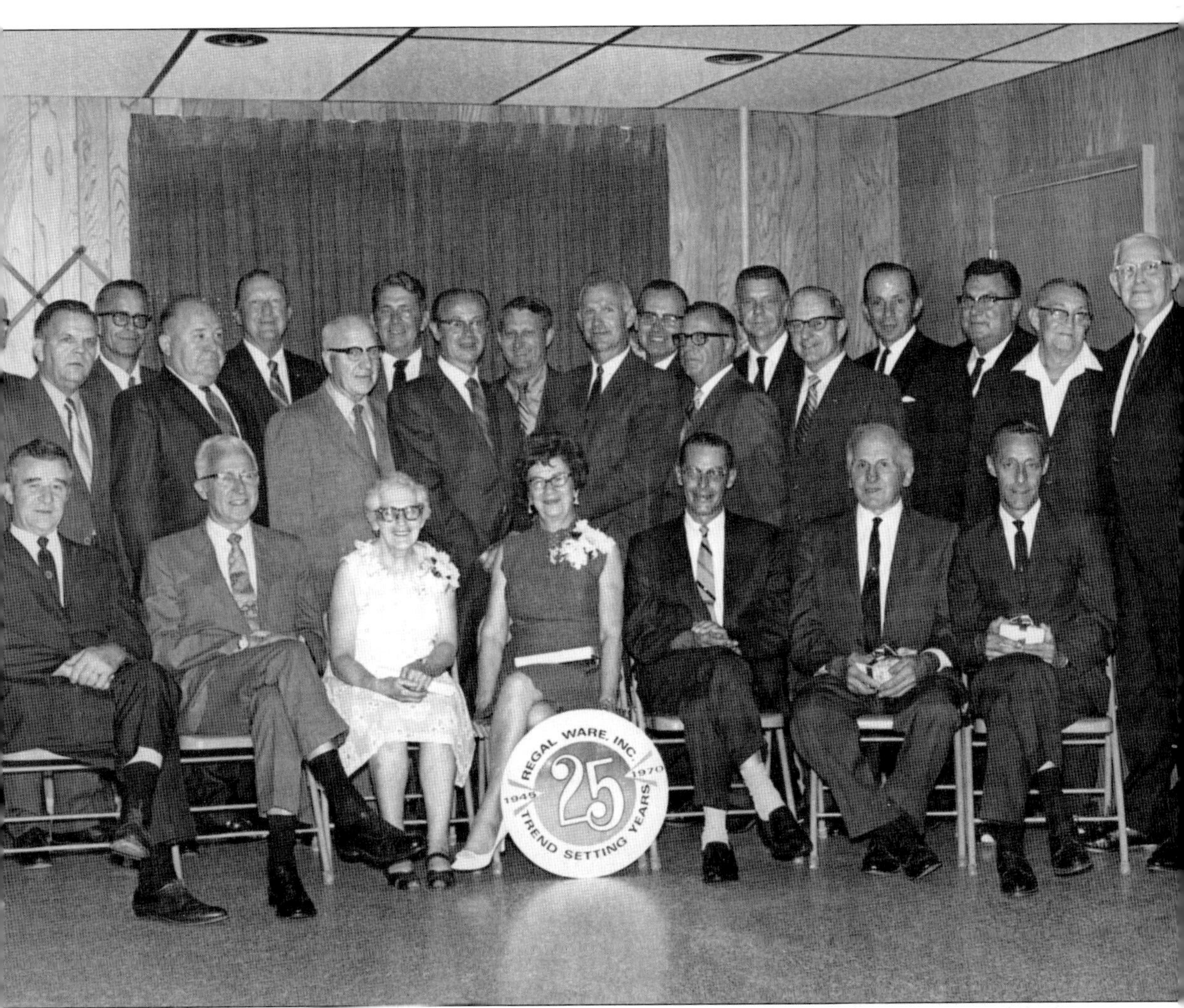

Regal Ware Inc.'s first 25 Year Club celebration was in 1970. Club members are Christopher Kober, Ludwig Roecker, Eldin Ebert, Roman Kral, Milton Meister, Marvin Martin, William McCarty, Lloyd Hron, George Koerble, Frederick Raether, Milton Maedke, Albert Hron Jr., George Eggert, Ralph Remmel, Harold Krueger, Leroy Keller, Arnold Ziemet, Harold Meisenheimer, Alfred Kral, L. N. Peterson, Bertha Ebert, Edna Oster, Lyle Manthei, John Diels, and Willard Manthei.

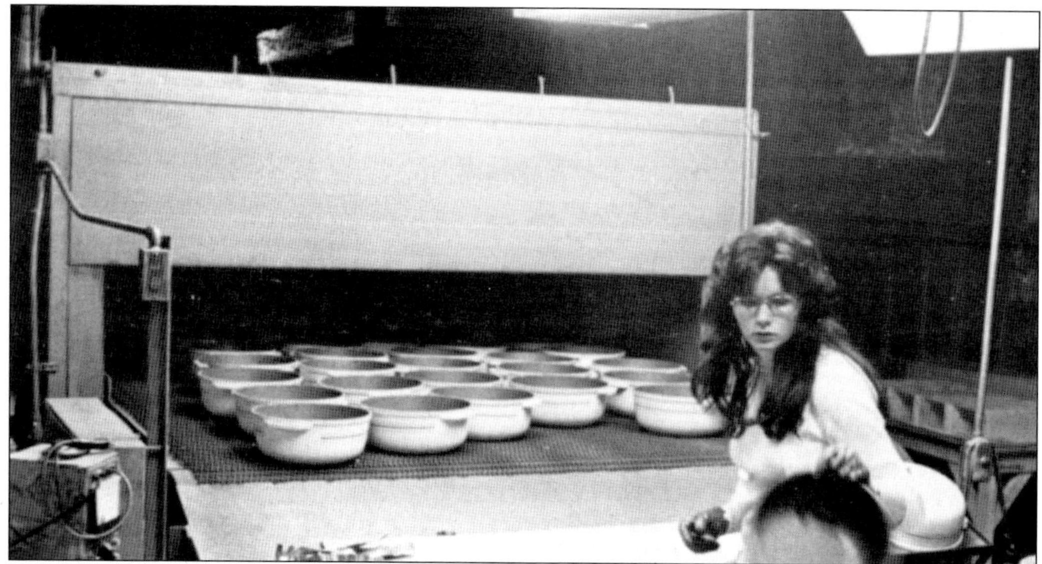

At Regal Ware Inc. nonstick coating was applied automatically in the 1970s. Trained employees applied coatings such as DuPont's Teflon II and Silverstone, which are seen coming off of the line in this photograph.

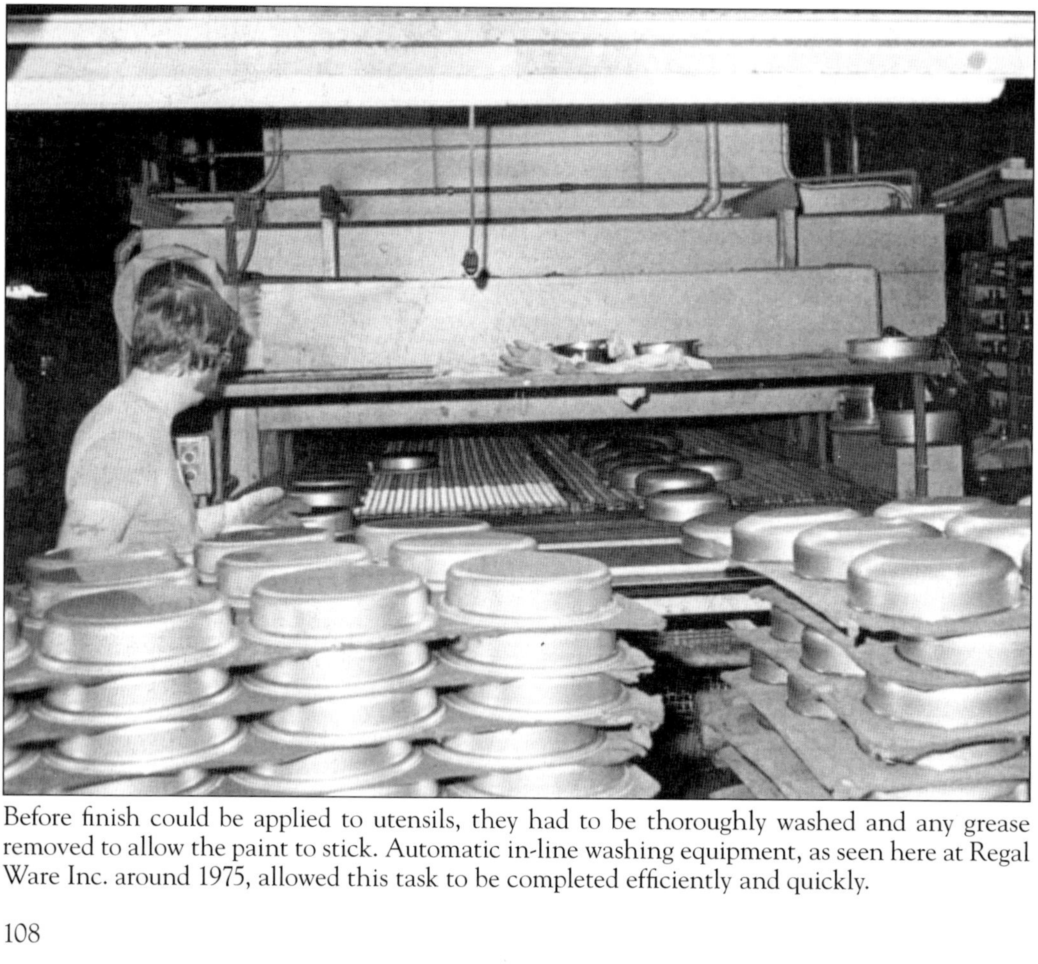

Before finish could be applied to utensils, they had to be thoroughly washed and any grease removed to allow the paint to stick. Automatic in-line washing equipment, as seen here at Regal Ware Inc. around 1975, allowed this task to be completed efficiently and quickly.

One way to finish cookware was polishing or buffing. The utensil could be made to shine with a soft buff and polish, or could be brushed with a wire attachment to create a softer finish. Here a Regal Ware Inc. employee polishes a pan around 1975.

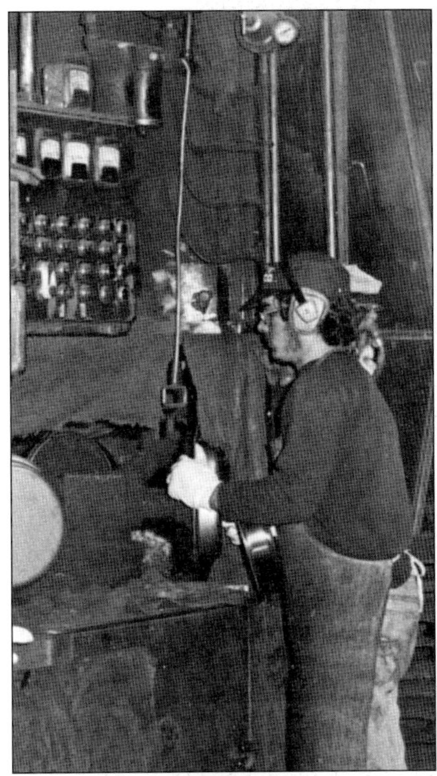

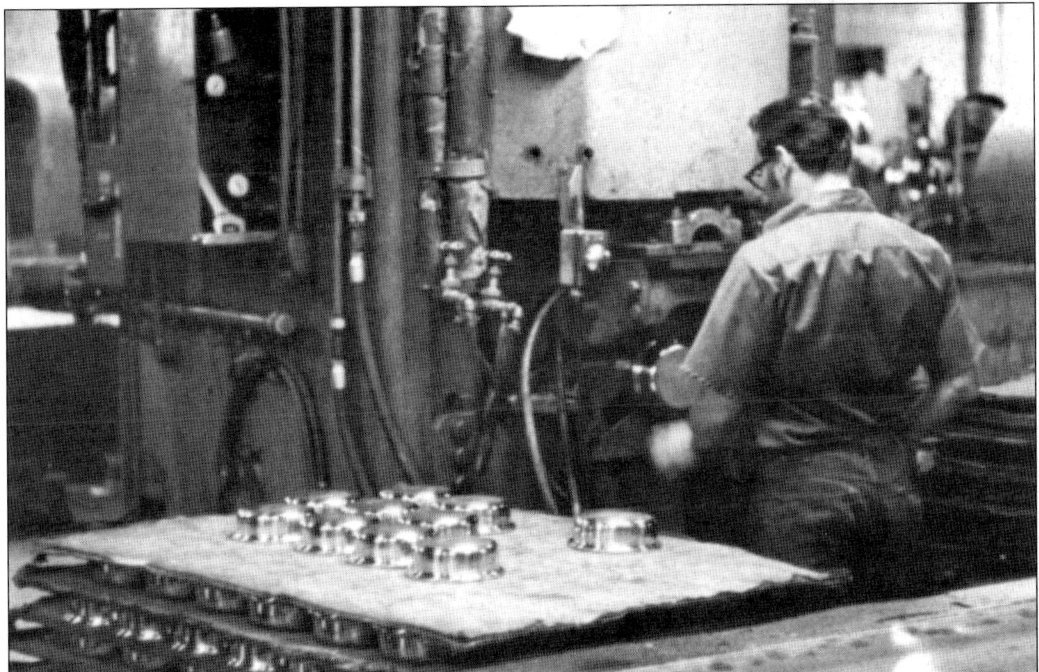

Regal Ware Inc.'s spot welder Richard Ebert is working on the finishing line of stainless steel cookware. The finishing line was where the final touches were put on the utensils. Handles and knobs were attached, a final inspection, and packing were all part of this line.

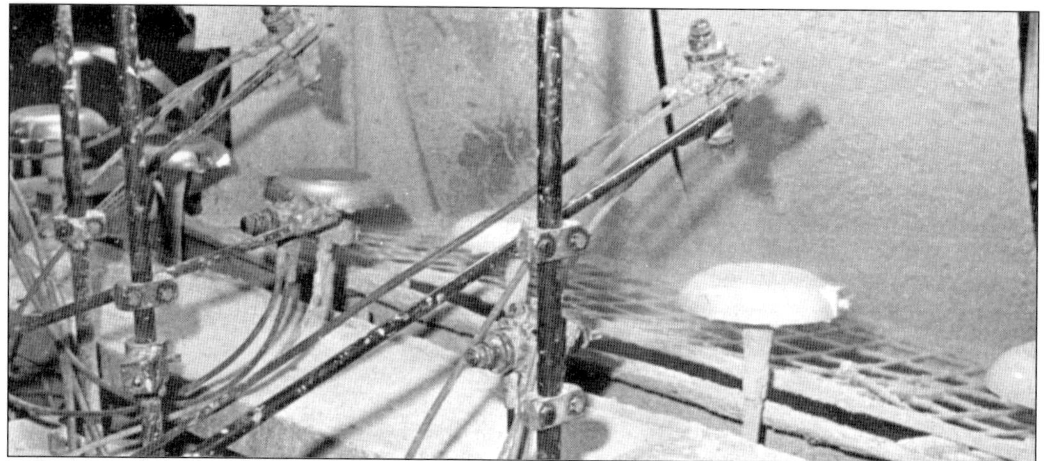

This photograph shows Regal Ware Inc.'s automated painting and silk-screening process. Porcelain enamel, baked enamel finishes, and silk-screen designs were applied before being baked in an oven. Here the utensil is held on a support while the painting arm sprays the finish.

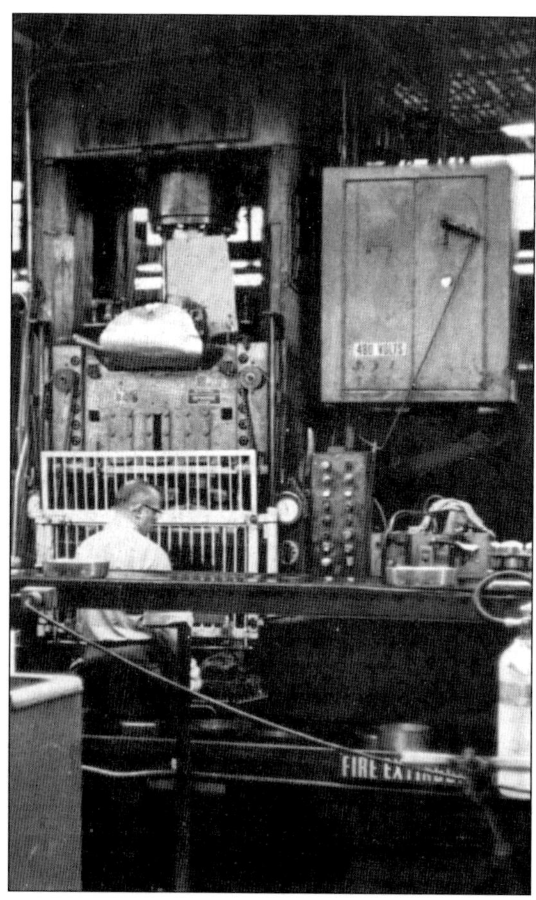

Here Harvey Jandre operates one of Regal Ware Inc.'s draw presses around 1970. The white bars in front of Jandre are a safety gate on the front of the press.

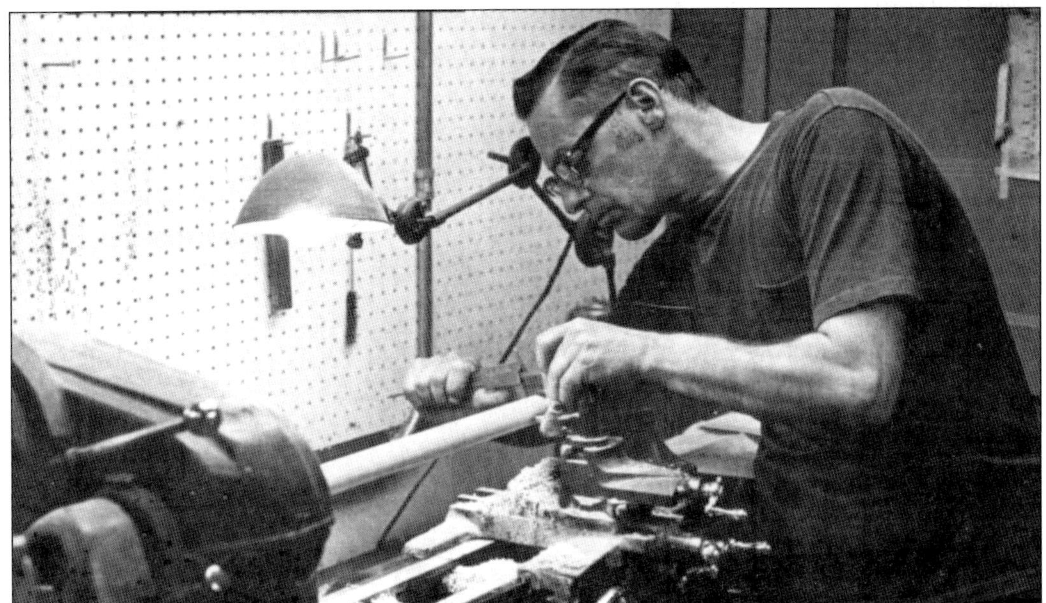

Lyle Manthei carves a wooden handle for a model product in Regal Ware Inc.'s model shop. Prototypes were made of all products to check for design and engineering flaws before they were put into production. This photograph is from around 1970.

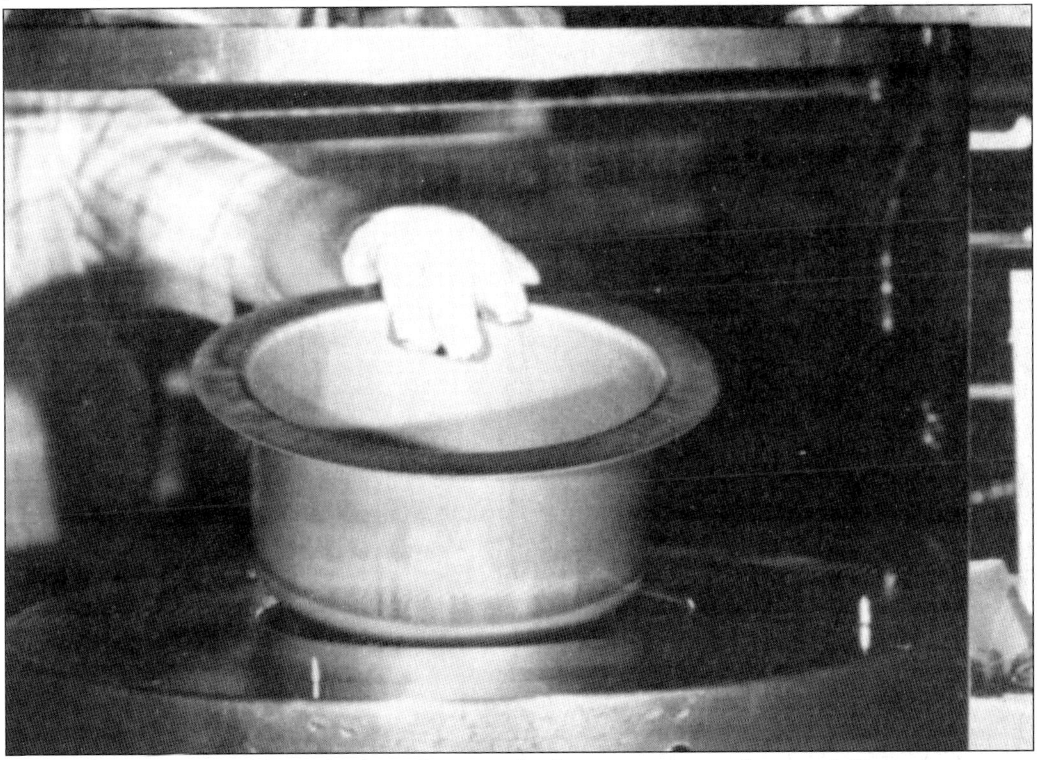

Regal Ware Inc. was able to do deep draws with aluminum or stainless steel. Deep draws were used to create the large, deep pans and large coffee urns. This photograph shows one of the deep draws being removed from the press in the 1970s.

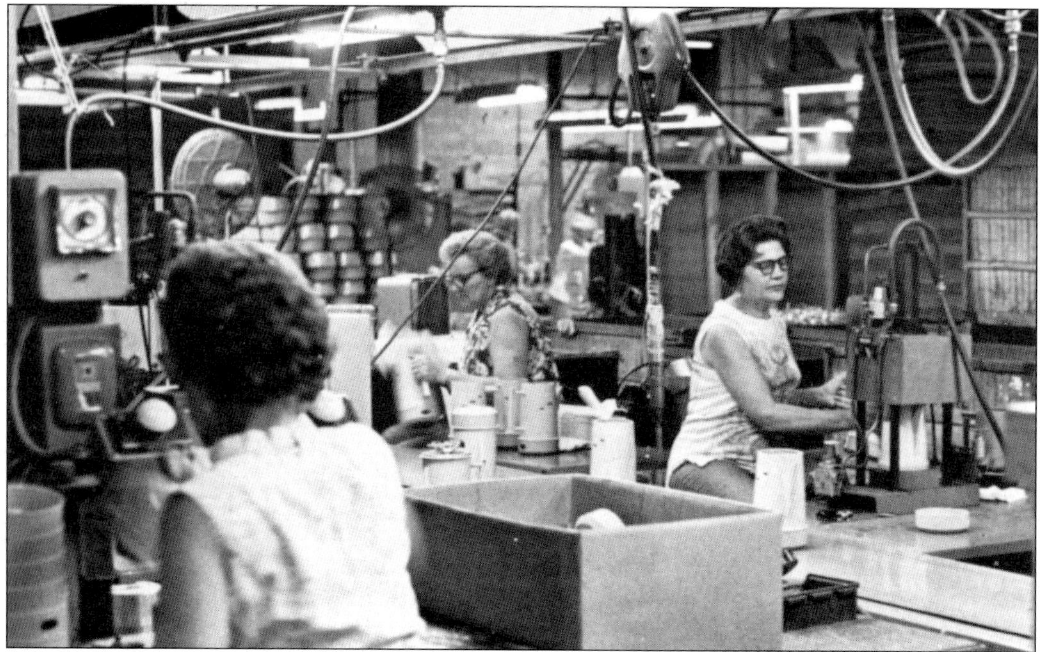

From left to right, Mary Bruessel, Virginia Wendorff, and Hilda Govin work in the Electrical Assembly Department at Regal Ware Inc. This department was responsible for the final assembly of all electrical appliances. In this photograph, from around 1975, Poly Perk coffeemakers are being assembled.

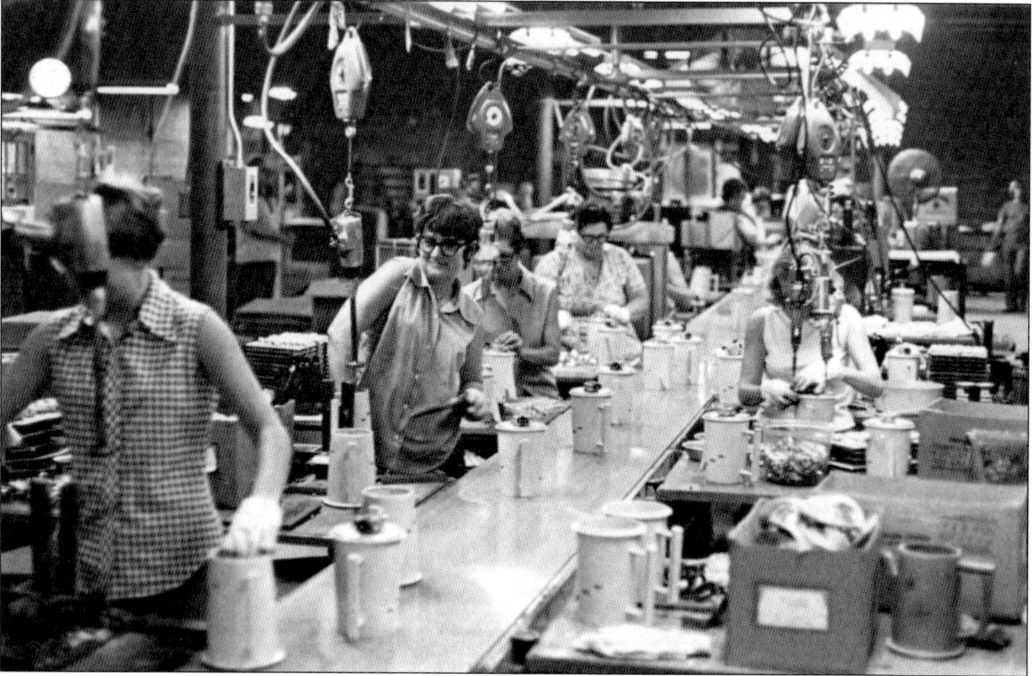

Poly Perk coffeemakers are being assembled around 1975. On the left side of the line, from front to back, are Donna La Shay, Phyllis Faber, Marlea Miller, and Irene Brandenburg. On the right is Johanna Gavin. The Poly Perk was one of Regal Ware Inc.'s best-selling products.

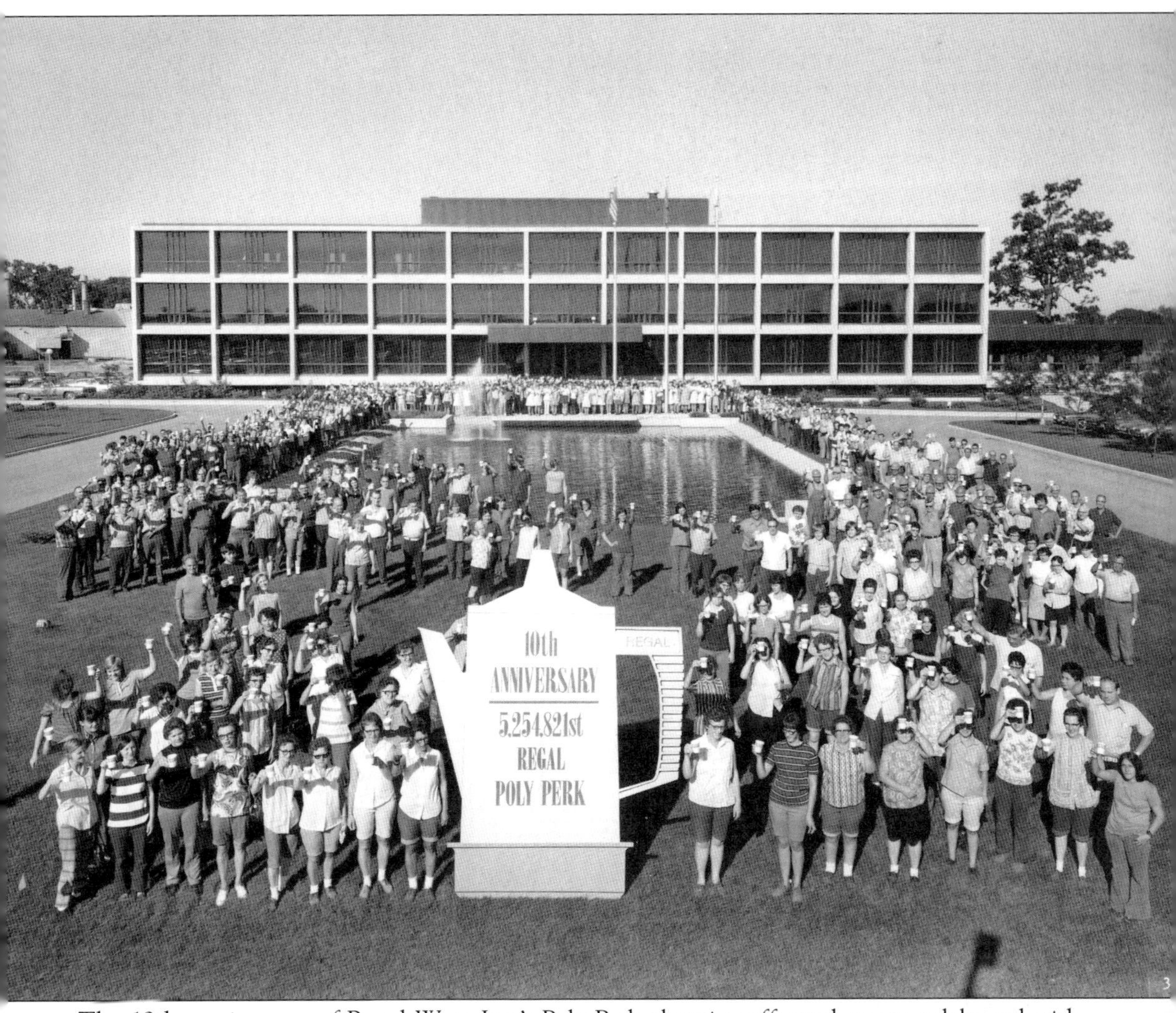

The 10th anniversary of Regal Ware Inc.'s Poly Perk electric coffeemaker was celebrated with a coffee break in front of the company's office building on September 5, 1973. The Poly Perk was one of Regal Ware Inc.'s best-selling products and was the best-selling coffeemakers in the housewares industry. The coffeemaker was made with injected mold technology. Although the initial outlay for the equipment was expensive, it allowed minor changes to design and color to be made inexpensively, thus allowing many different products from the technology.

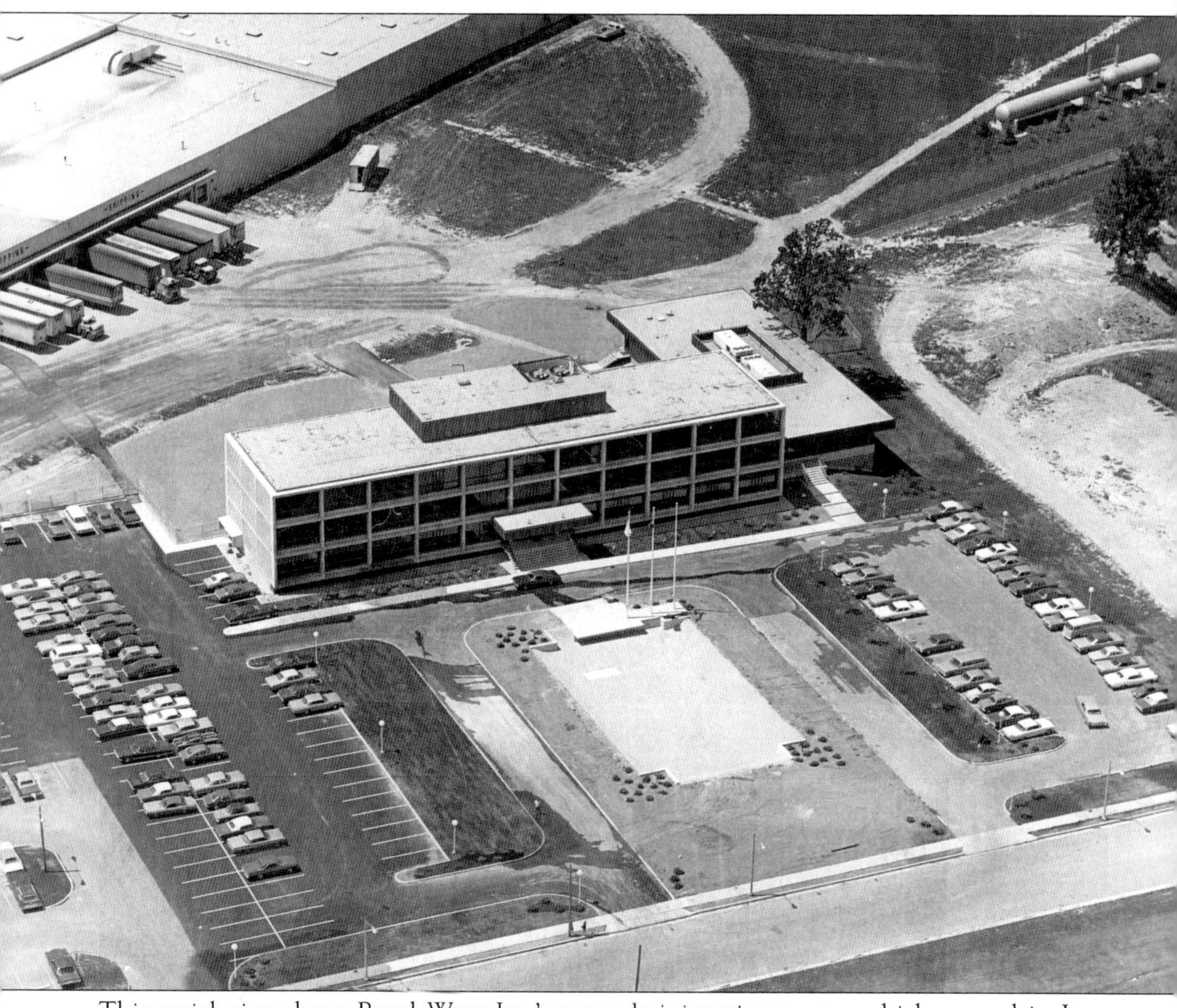

This aerial view shows Regal Ware Inc.'s new administration center, which opened in June 1971 in Kewaskum. The new four-story complex centralized the management in one location for improved operations. Designed by Architects III of Milwaukee, the design was meant to complement the Kettle Moraine landscape and featured a modern office flow design.

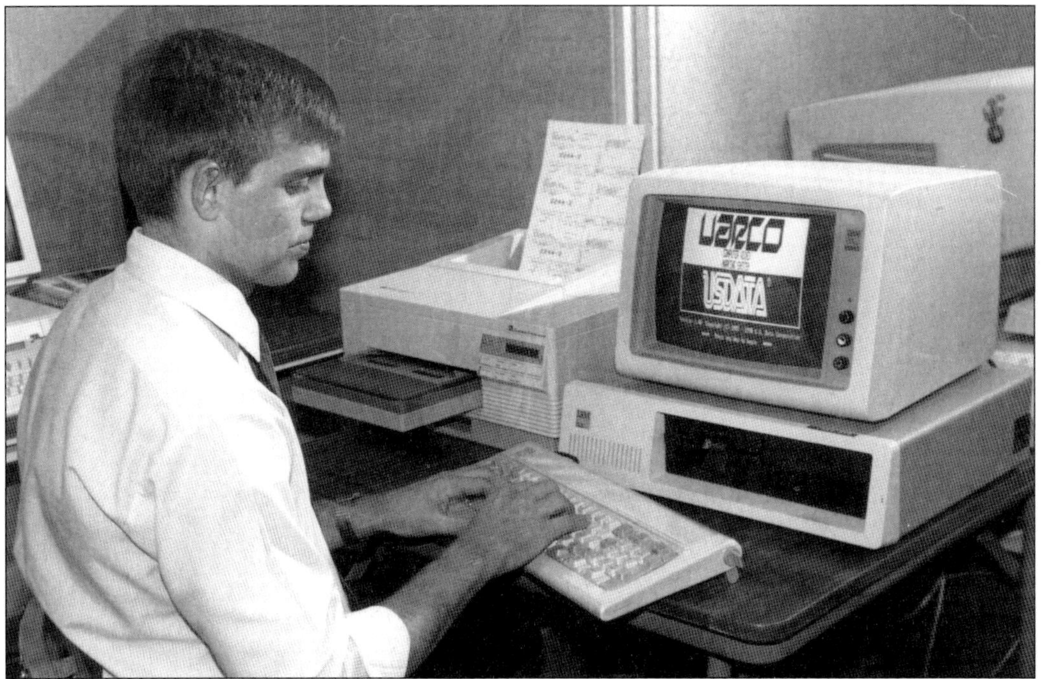

A Regal Ware Inc. employee demonstrates the latest technology in customer ordering at the company's open house in October 1971. Computerized data speeds order processing and distribution.

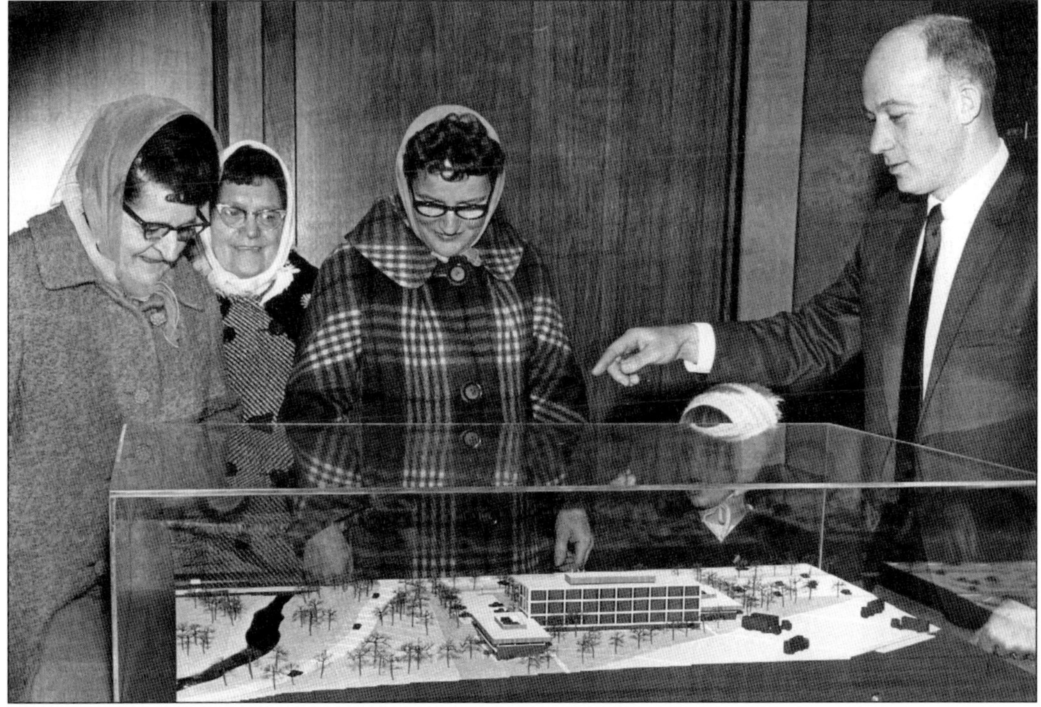

A group of visitors look at a model of Regal Ware Inc.'s new administration center. The observers are attending an open house in October 1971, celebrating the opening of the new building.

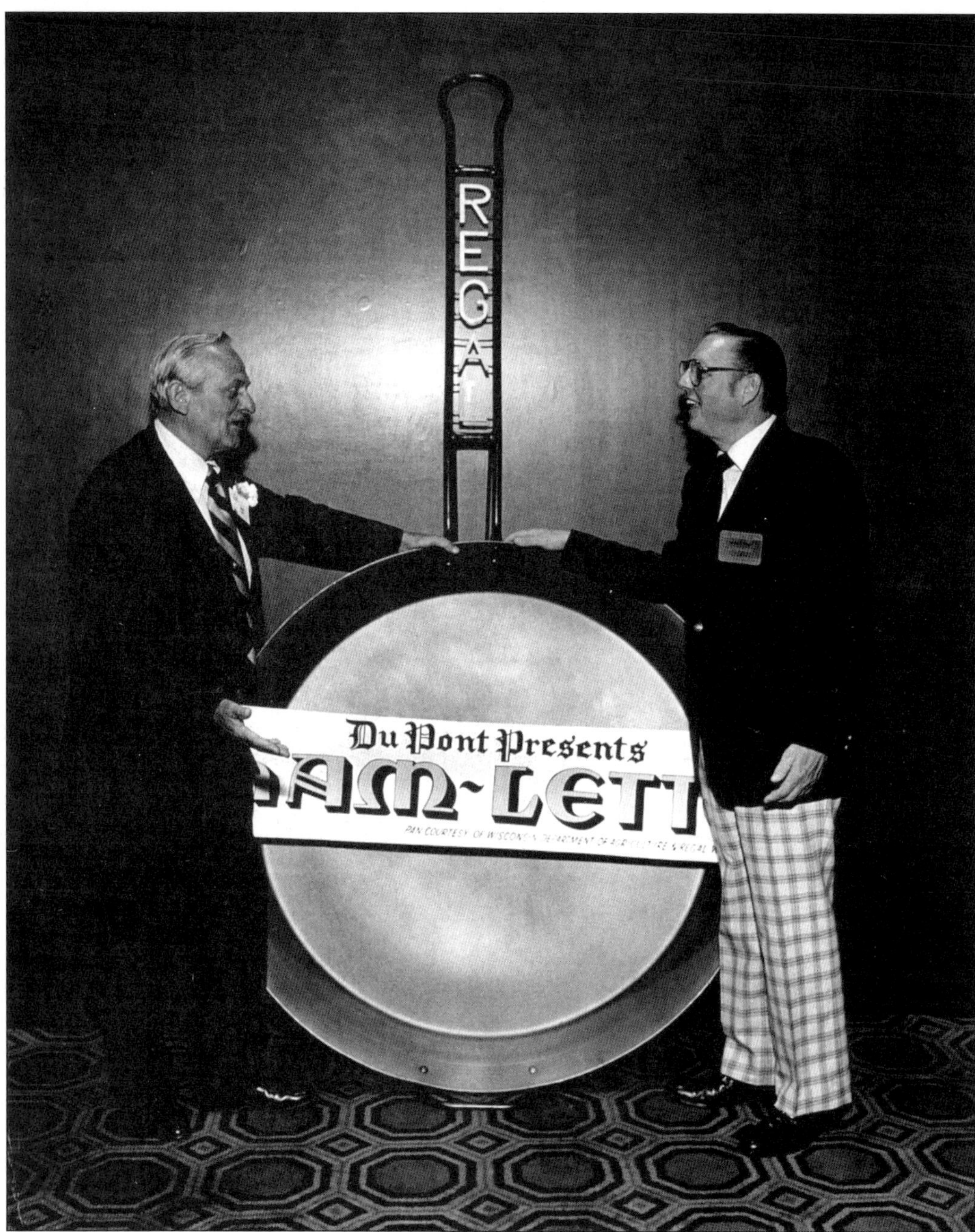

David Lee on the left and James D. Reigle show off the 54-inch skillet that was manufactured by Regal Ware Inc. It was coated with Silverstone, a nonstick coating created by DuPont Teflon, introduced on cookware in 1961. It was the world's largest nonstick skillet. Made for the Wisconsin Department of Agriculture, it was used at county fairs and agriculture exhibits to prepare 300-egg omelets, enough to serve 800 people. The pan also served in displays at trade shows.

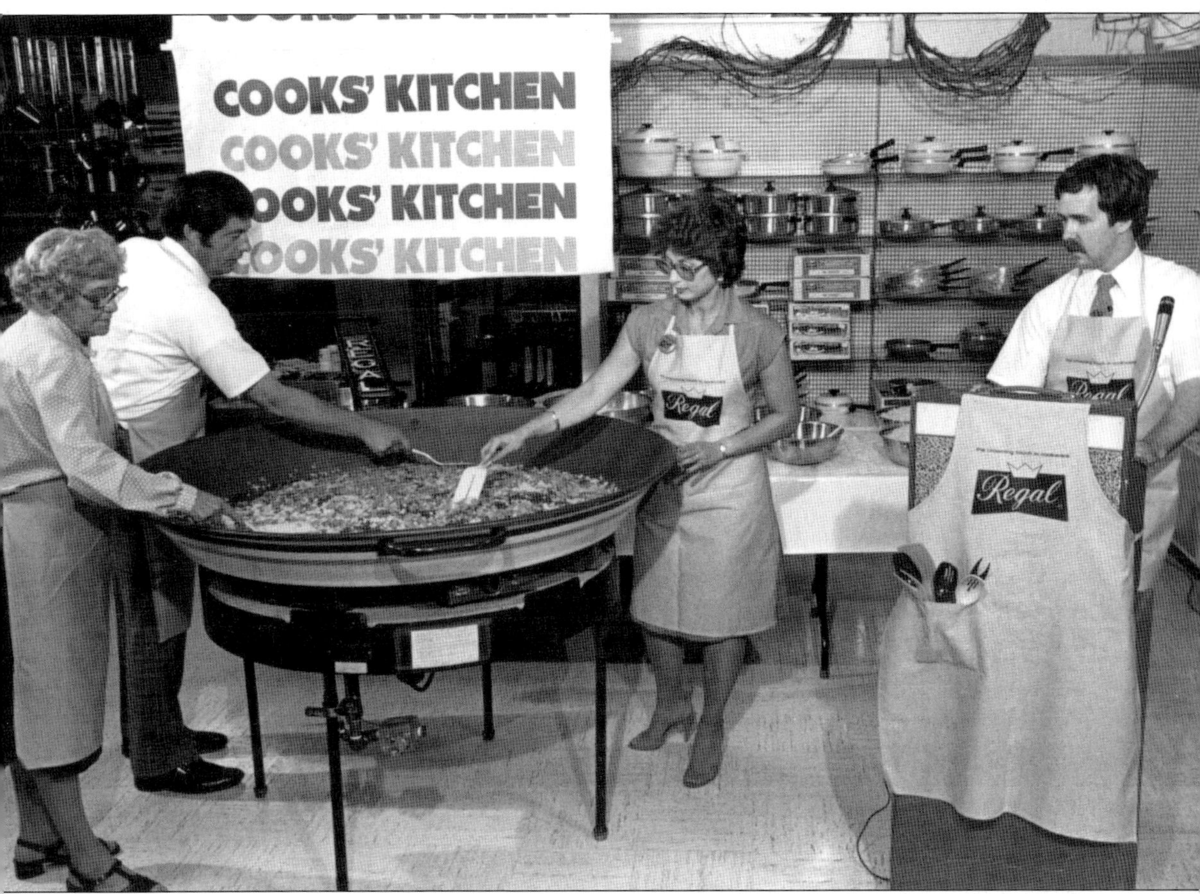

This photograph shows Regal Ware Inc.'s giant fry pan in operation in the 1980s. The pan was used across the county as a promotional tool for Regal Ware Inc. cookware and as a general customer draw for the store or event. Retailers wishing to use the pan in a promotion were charged a small fee to defray transportation and food costs. Here from left to right, Jean Winkler, Jim Dorn, Mona Marino, and Rick Stroo demonstrate the nonstick properties of Regal Ware by cooking scrambled eggs in a giant sized portion. The recipe for huevos a la Mexicana called for 300 eggs, 4 pounds of butter, 40 tomatoes, and 10 pounds of cheese.

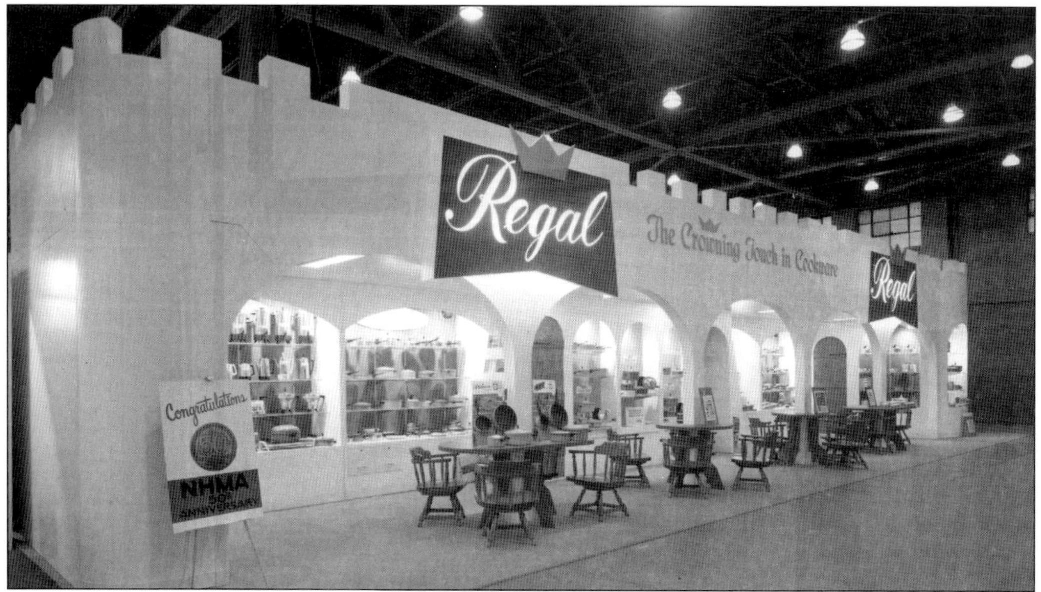

By 1989, the booths at the National Housewares Manufactures Association annual Cookware Show in Chicago had become more elaborate. Companies went all out for the 50th anniversary of the show, and Regal Ware Inc. was no exception with its stylized castle. In the past the show had been held twice annually, but is currently held once a year. Several other shows were also very important to the industry. Premium shows, grocery shows, and direct sales shows are just of few of these.

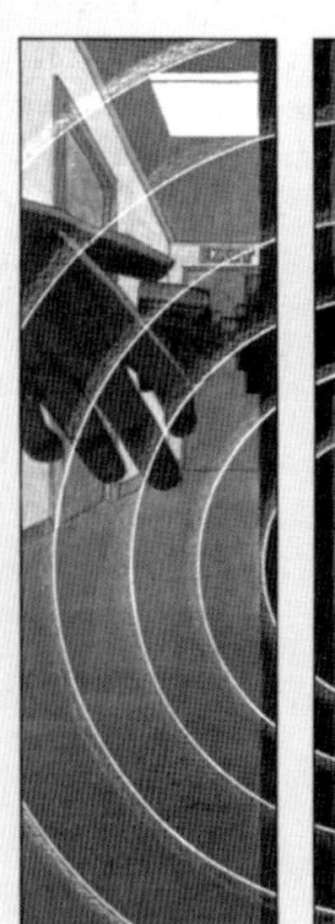

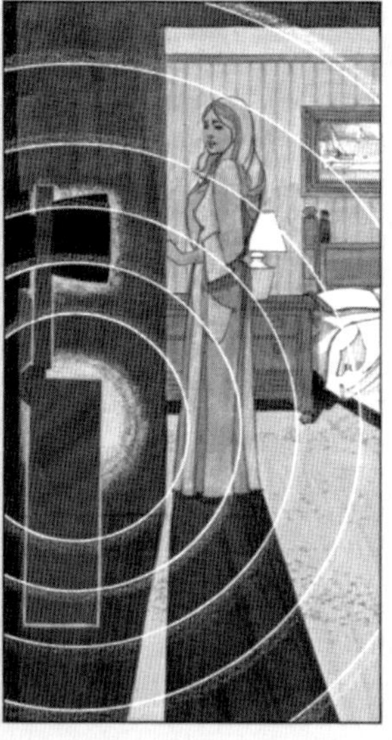

REGAL STARTLER®
PORTABLE ELECTRONIC
BURGLAR ALARM

Solid state electronic security system works on all wood and non-metallic doors in homes, apartments, hotels and motels. Hang Startler on inside doorknob. Piercing alarm signals when would-be-intruder touches the doorknob. Self-contained unit features sturdy molded plastic construction and sensitivity control. Requires one 9-volt alkaline battery (not included).

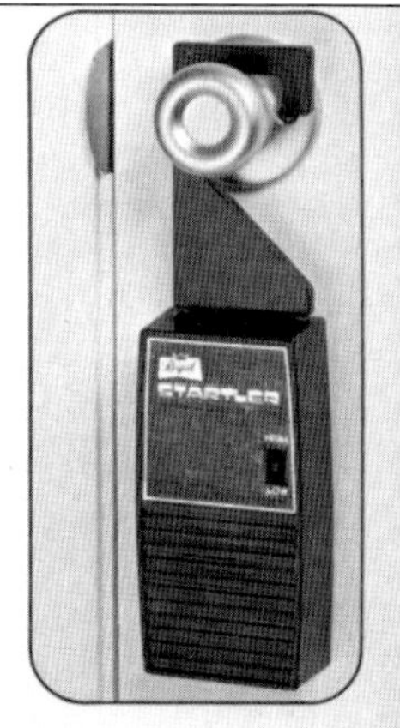

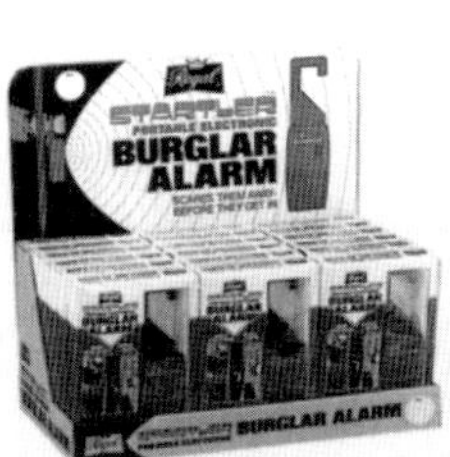

No. K6700 Startler Portable Burglar Alarm. Individual full-color window carton; repacked 6 to a master; Shipping Wt. 6 lbs.

No. 6708 Startler Display. Individual full-color window carton; prepacked 18 in full-color shipper/display carton; Shipping Wt. 19 lbs.

The Startler Burglar Alarm was produced by Regal Ware Inc. from around 1978 to 1984. The Startler is a wonderful example of the diverse products the company branched out into. This item was hung on a door knob and created an alarm sound when the knob was touched and the circuit was completed.

Four

The Transition Years 1980–2005

As the economy declined in the 1980s, consumers began to favor low price over quality of goods, and foreign products gained a hold in the market. To maintain competitiveness, many U.S. companies were forced to move manufacturing overseas, began automating factories, and abandoned prominent product lines focusing on select housewares products.

Baby boomers were once the consumers of millions of small electric appliances and cookware. When this group's focus shifted to saving money instead of spending, manufacturers were challenged to appeal to a younger market with disposable income. Unlike baby boomers, the younger market was interested in electronic gadgets and entertainment rather than housewares. These consumers worked full-time jobs and cooked less but wanted fully-equipped kitchens with all of the trendy utensils. Unlike the previous generation, cookware became a status symbol rather than a necessity of convenience. Advertisers attracted new consumers through the interest in health and wellness products. Water purifiers and waterless nonstick cookware limited cholesterol, fat, and salt in the diet. Small electric appliances, once a staple of the housewares industry, took a back seat to the reemergence of cookware. Cookware has been given a new life as cooking has become a hobby instead of a necessity for many Americans.

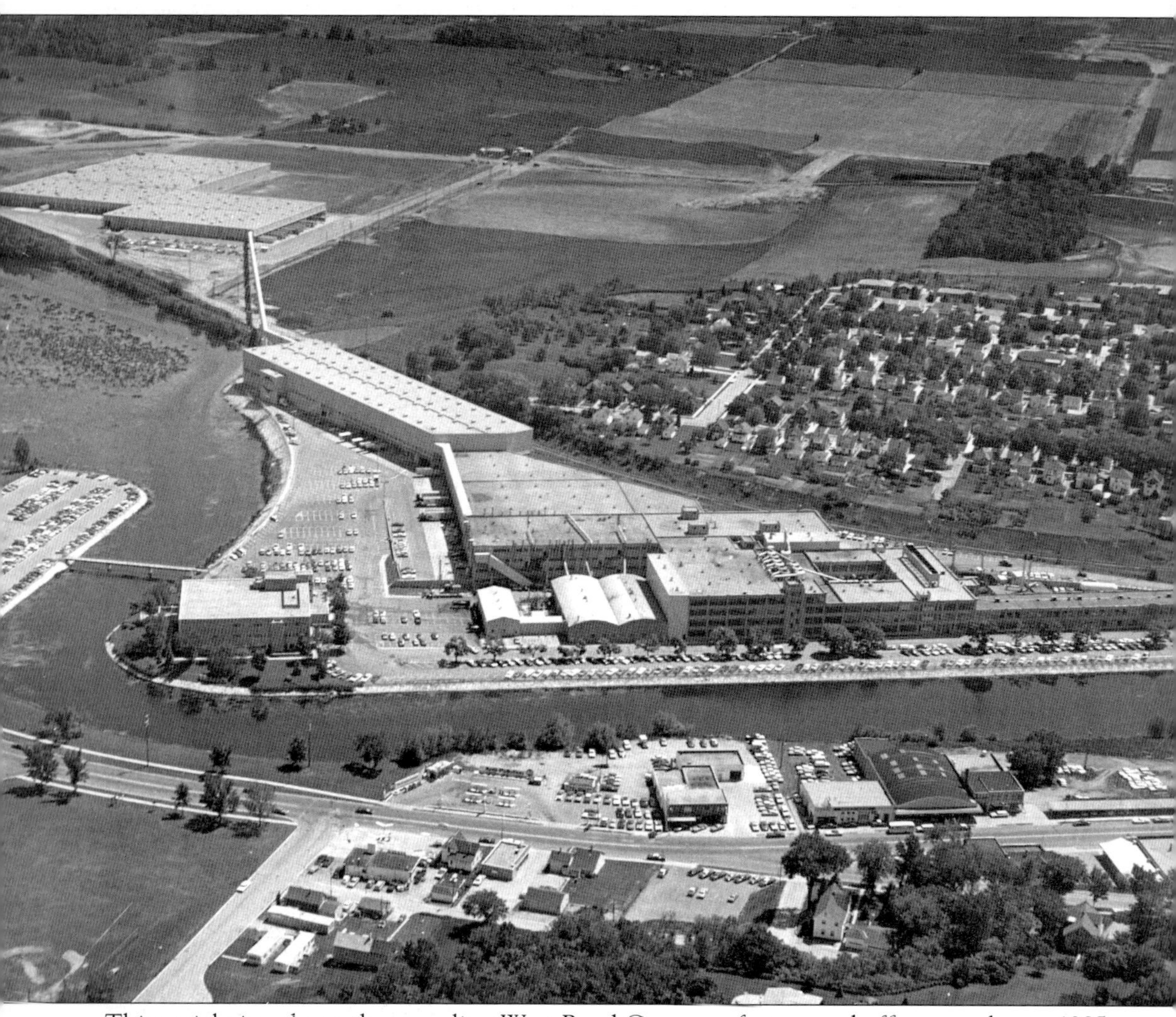

This aerial view shows the sprawling West Bend Company factory and office complex in 1985. A conveyor belt connected the building in the foreground with the one in the background. The original three-story building can be seen in the front on the right, wedged between a four-story and two-story addition. During the 1980s and 1990s, the West Bend Company went though many changes of ownership. After its merger with Rexall Drug and Chemical Company in 1968, the companies reorganized into Dart Industries. In 1980, Dart Industries merged with Kraft Inc. becoming Dart and Kraft Inc. In 1986, West Bend became part of Premark International Inc., which split from Dart and Kraft Inc. In 1999, Illinois Tool Works Purchased West Bend from Premark International Inc. before finally selling it to Regal Ware Inc. in 2002.

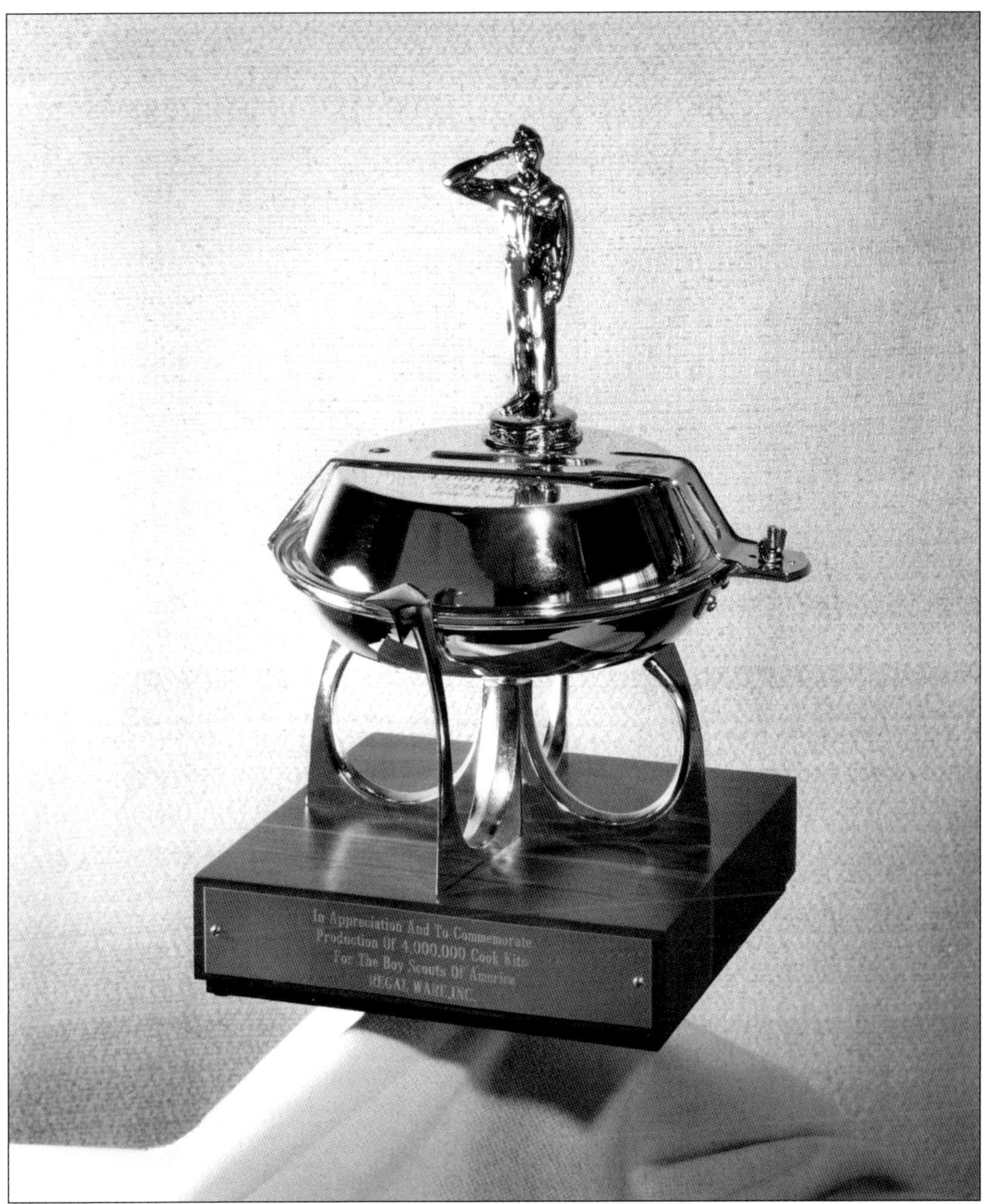

Regal Ware Inc. began producing cook kits and canteens for the Boy Scouts of America in 1948. In 1980, the Boy Scouts of America presented their Award for Excellence to Regal Ware Inc. in recognition of the company's service to the Scouts. With over 700 suppliers to the Boy Scouts, Regal Ware Inc. was one of only five to receive the award. Regal Ware Inc. was chosen on the basis of high quality, fast delivery time, ease of processing, good communication, and general cooperation with the Boy Scouts of America.

Regal Ware Sells West Bend
Retail Housewares Division

7/24/03

Kewaskum manufacturer egal Ware, Inc. has sold its Vest Bend® Retail lousewares Division to U.S.-ased Focus Products Group, LC of Vernon Hills, Illinois ccording to documents igned by executives from oth companies Tuesday.

The West Bend® Retail lousewares Division has een marketing small electri-al appliances through estab-ishments such as depart- approvals.

Offices, warehousing and distribution of West Bend® Housewares products will remain at its current location of 1100 Schmidt Road in West Bend, Wisconsin.

"We are impressed with the high quality of the West Bend Retail Housewares organiza-tion and products, and are committed to strengthening the West Bend® brand name in the retail industry," said November 2002. "The sale of the Retail Housewares unit will not affect the Direct Sales Cookware or Water Systems businesses," according to Regal Ware President and CEO Jeffrey Reigle. "This move is consistent with our strategic direction to focus on the Cookware and Water Treatment direct sales arena, while maintaining our growth position in sales to Commercial businesses such United States, Can Mexico, Asia and Europe, include such quality item range-top cookware, c mercial coffee urns, home drinking water tr ment systems.

Focus Products Group rently has two wholly ow subsidiaries, AM Houseworks, LLC Sensible Storage, LLC. AM sells kitchen utensils to cialty retailers and dep

In 2002, Regal Ware Inc. acquired selected assets of the West Bend Company from then owner Illinois Tool Works. The assets included the West Bend Company East Plant on Schimidt Road in West Bend, machinery, furniture, brands, trade names, consumer goodwill, distribution network, and product inventory. In 2003, Focus Products Group LLC acquired the business and operations of the West Bend Company Retail Housewares Division from Regal Ware Inc. This newspaper clipping from the July 24, 2003, *West Bend News* tells of the later transaction. Most of the main West Bend factory site has been converted to high-rise condos and apartments.

James D. Reigle spent time in the U.S. Army Air Corps before beginning his career in the Shipping and Receiving Department of the Kewaskum Utensil Company in 1949. After the retirement of his father and company founder J. O. Reigle in 1965, James D. took over as president of Regal Ware Inc. Although the third generation of Reigles took over as president in 1992, James D. continues to serve as chairman of the board and serves on the company's executive committee.

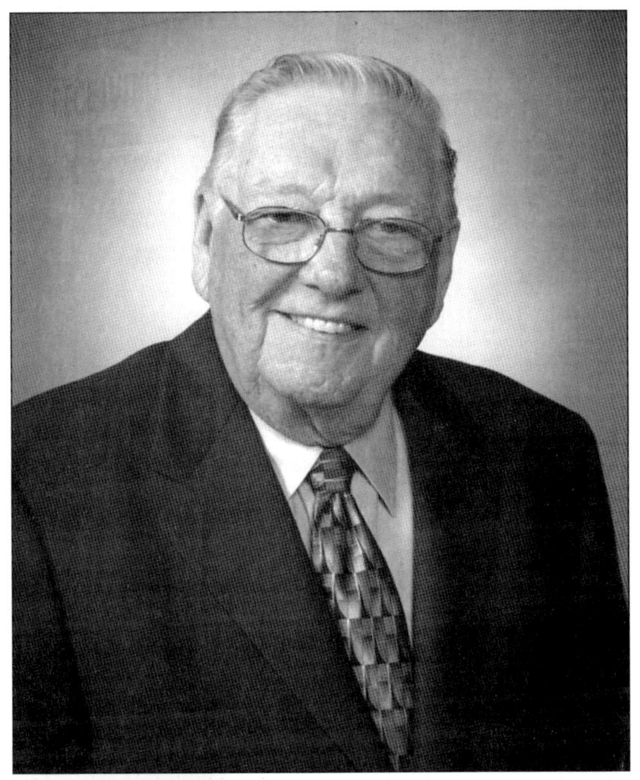

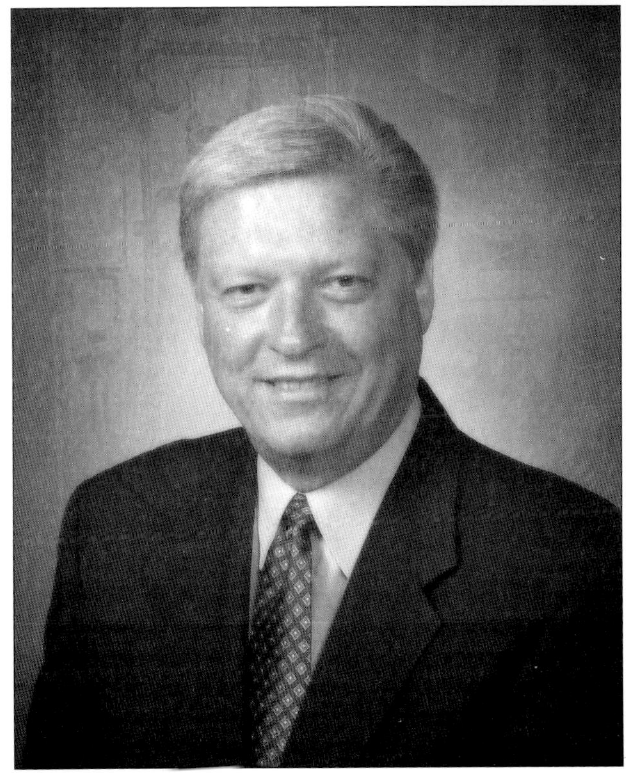

Jeffery A. Reigle is the current chief executive officer of Regal Ware Inc. and took over as president of the company in 1992. He has worked for the company since 1973, when he joined as a management trainee after completing a degree in business. After gaining experience in sales and production control he joined the Regal Ware Inc. Board of Directors in 1979. He is the third generation of the Reigle family to hold the position as chief executive officer.

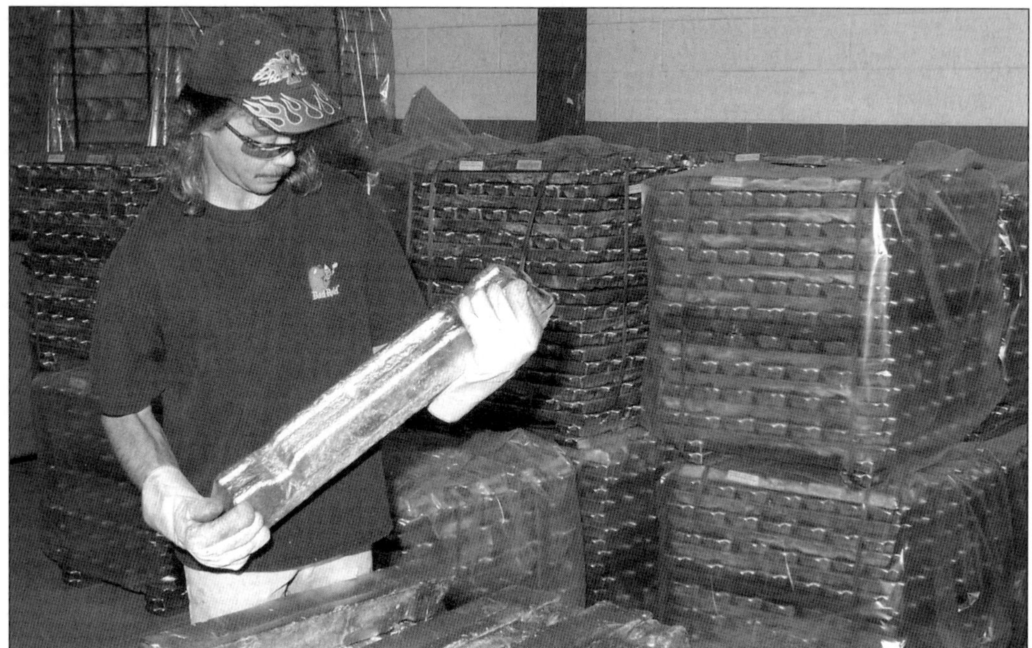

Regal Ware Inc. continues to produce quality cookware, including cast aluminum. This photograph shows employee Rodney Kutz examining an aluminum ingot for impurities such as oxides or dirt. Aluminum ingots are obtained from various plants throughout the United States.

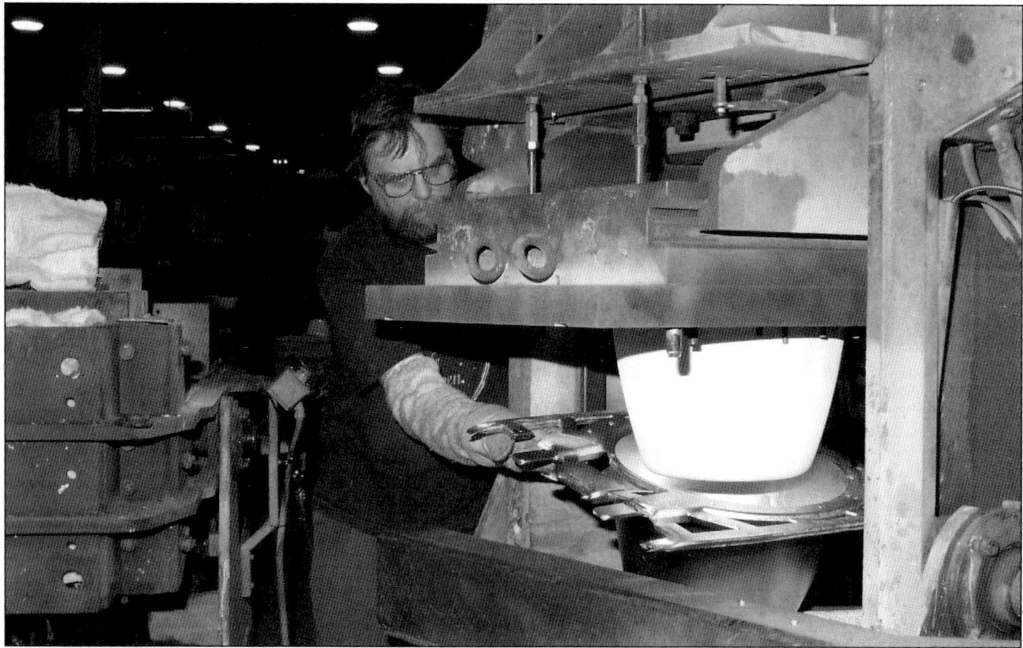

After being melted, the aluminum is poured into a mold. Here Gary Kamrath removes a cast aluminum product from its mold. The utensil is still very hot, and heat resistant gloves are required. The piece will then be let to cool and then inspected. Excess aluminum is then removed from the edges of the pan. Any utensils rejected because of imperfections are recast so there is very little waste.

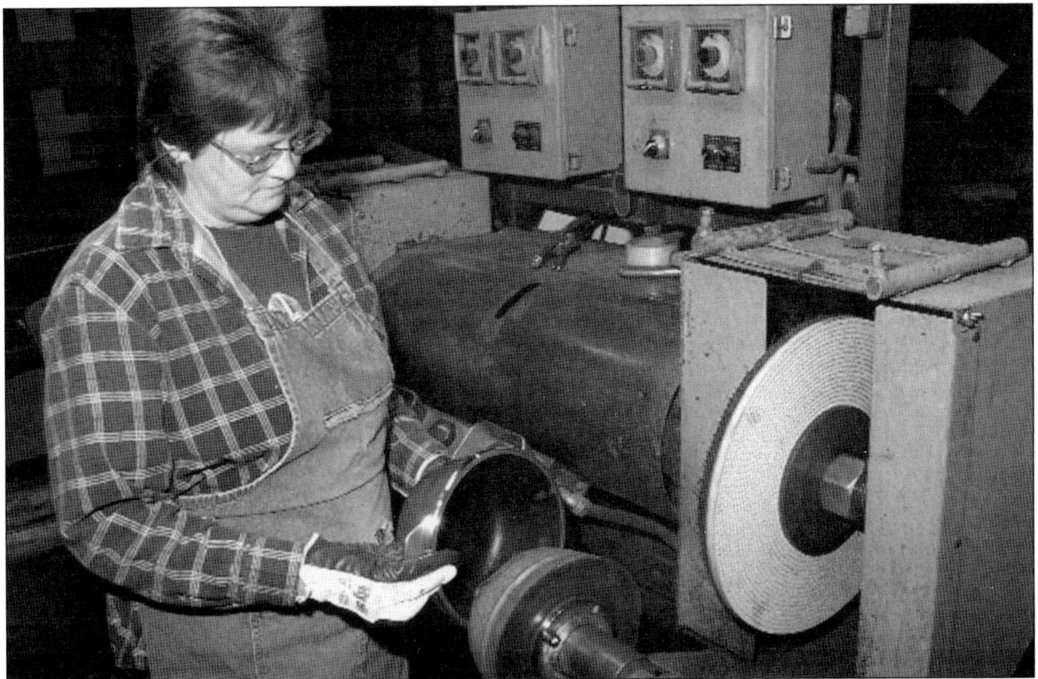

One of the final steps in processing cast aluminum cookware is buffing. Bonnie Burmeister is seen here buffing the top edge of the cast aluminum pan to a high polish finish. Different finishes are achieved with different types of buffing and polishing brushes.

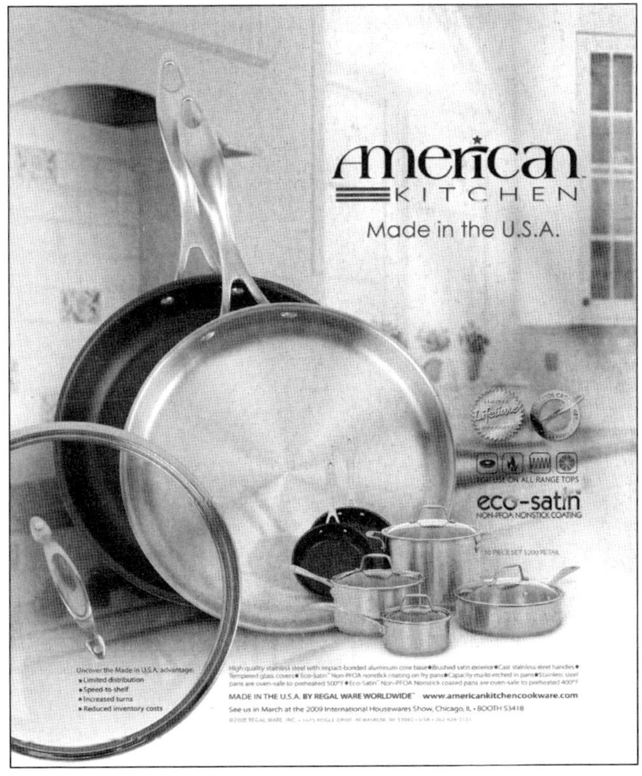

Regal Ware Inc. continues its innovations with American Kitchen, which was introduced in March 2009 by the company. The utensils are made of stainless steel with an encapsulated aluminum core bottom. The product was designed for functionality as well as beauty and can go from stove top to oven to table. Innovations include eco-friendly nonstick interior coating and it is able to be used on traditional as well as induction ranges. Through the years, Regal Ware Inc. made manufacturing in the United States a priority. While a small amount of work was outsourced, in 2008, Regal Ware Inc. returned all cookware production back to the United States. This move was made to capitalize on faster distribution and better control over quality.

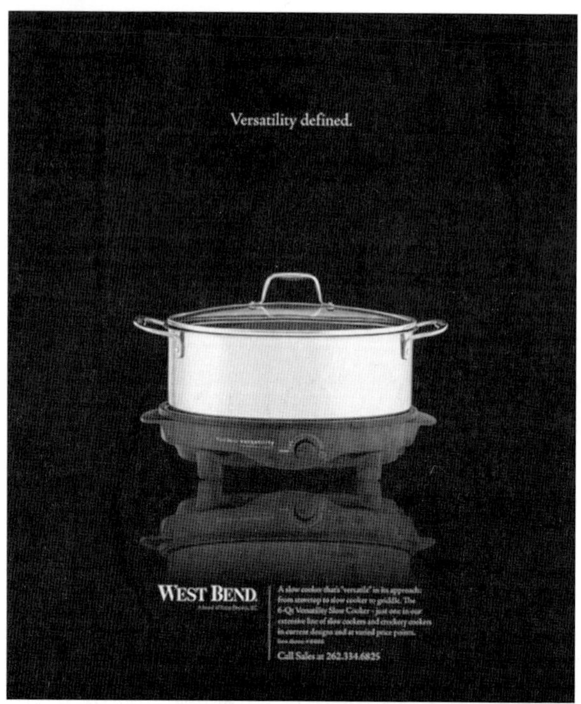

This 2008 advertisement is for one of the many products still produced under the West Bend name. Focus Products Group LLC continues the tradition of quality with products such as the Versatility Slow Cooker. This particular product was sold through Ace Hardware and Amazon.com. One of the main features of modern products is versatility. Aptly named, the Versatility Slow Cooker can go from the stove top, to heating base, oven, tabletop, refrigerator, or dishwasher.

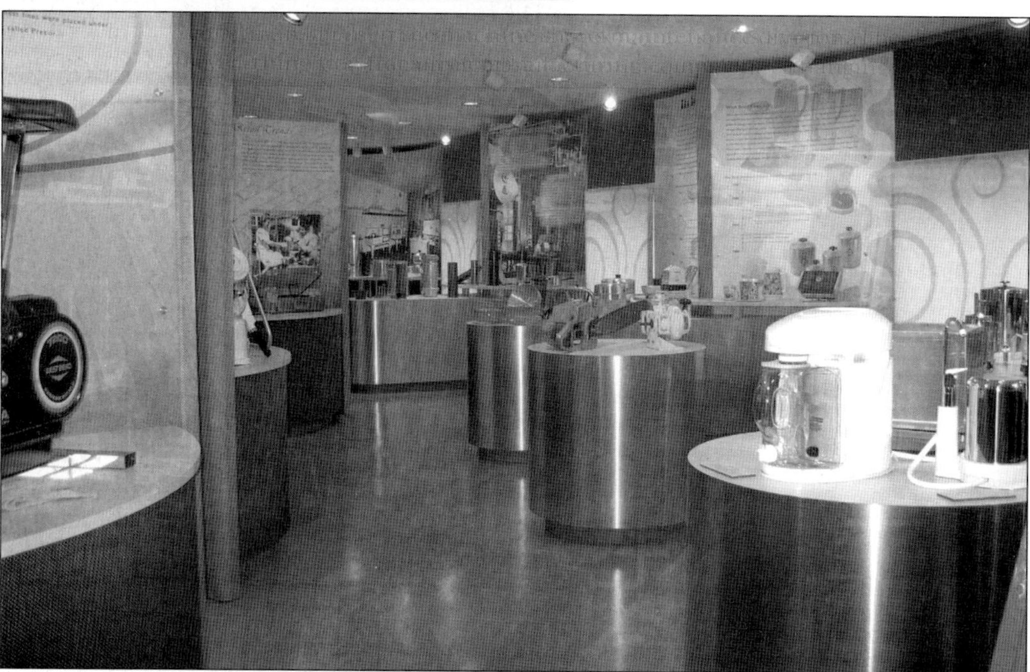

The West Bend Company/Regal Ware Museum was opened in November 2008. The museum, a site run by the Washington County Historical Society, strives to preserve and illustrate the local and wider-reaching history of the West Bend Company and Regal Ware Inc. Here over 3,000 objects and thousands of photographs and archival pieces from both the West Bend Company and Regal Ware Inc. and Focus Products Groups LLC are cared for by the museum staff and volunteers.

About the Washington County Historical Society

Since 1873, the Washington County Historical Society has been the steward of the rich history of Washington County. The society has Captured the Past through the operation of four unique museums and historic sites: the Old Courthouse Museum (seen here), the Old Sheriff's Residence and Jail, the St. Agnes Historic Convent and School Complex, and the West Bend Company/ Regal Ware Museum. For over 130 years, the Washington County Historical Society has Captured the Past for future generations.

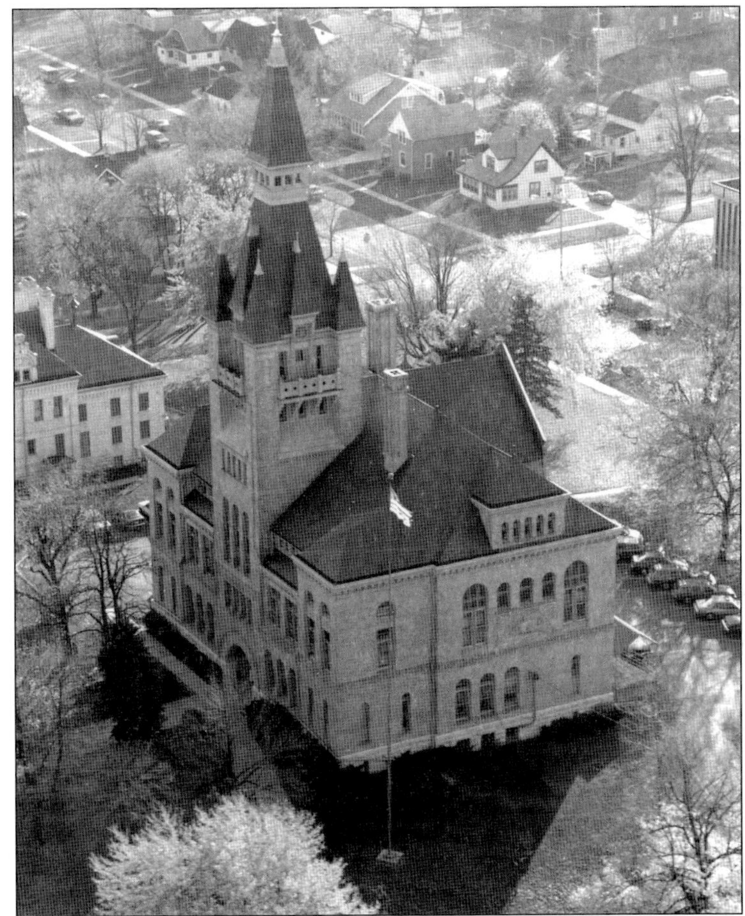

ACROSS AMERICA, PEOPLE ARE DISCOVERING SOMETHING WONDERFUL. *THEIR HERITAGE.*

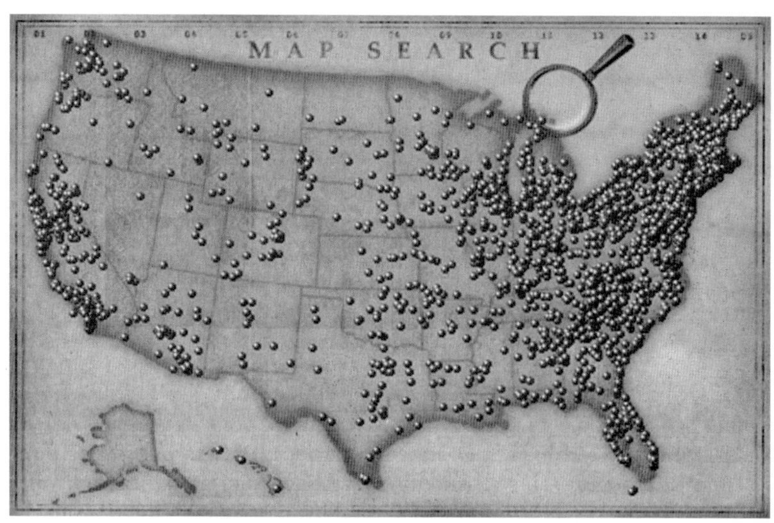